SPIN AND TORSION IN GRAVITATION

SPIN AND TORSION IN GRAVITATION

Venzo de Sabbata
Dept. of Physics, Bologna University, Italy;
Dept. of Physics, Ferrara University, Italy;
World Laboratory, Switzerland

C. Sivaram
Indian Institute of Astrophysics, India;
World Laboratory, Switzerland

World Scientific
Singapore • New Jersey • London • Hong Kong

Published by

World Scientific Publishing Co. Pte. Ltd.

5 Toh Tuck Link, Singapore 596224

USA office: 27 Warren Street, Suite 401-402, Hackensack, NJ 07601

UK office: 57 Shelton Street, Covent Garden, London WC2H 9HE

British Library Cataloguing-in-Publication Data
A catalogue record for this book is available from the British Library.

SPIN AND TORSION IN GRAVITATION

ISBN-13 978-981-02-1766-2
ISBN-10 981-02-1766-8

PREFACE

This book deals with spin and torsion in gravitation. The earlier on "Introduction to Gravitation" (World Scientific, 1985) gave a detailed exposition of the General Theory of Relativity developed by Einstein and others: in this book (of the year 1985) there is a chapter on the Einstein-Cartan theory and now we like to extend exhaustively this theory, that is the introduction of the spin in General Relativity, considering all possible physical consequences, spin being another universal attribute of matter besides the mass.

The spin of elementary particles was to play a profound role in atomic, nuclear and particle physics. The unity of special relativity and quantum mechanics through the Dirac equation led to spectacular new results in physics: the prediction of antiparticles and the intrinsic magnetic moment of elementary particles, to name e few. The spin of elementary particles manifested itself in several new effects in fundamental interactions such as splitting of nuclear energy levels and non-degeneracy of hadronic states in strong interactions, parity violation in weak interactions etc.
The fact that the effects of spin when considered in gravitational interactions can also lead to several interesting physical phenomena in both the micro and macro-world is not so well known. The literature is mostly confined to specialized articles read by only those few directly working on the subject. Even most physicists working on gravitation theory are not much aware of the interesting consequences of spin modified gravitational effects, especially those caused by torsion which is the geometric effect of spin in space-time (analogously to mass causing space-time curvature).

The book we are contemplating could fill this gap and give an exposition of both the old and new results of spin and torsion effects on gravitational interactions with implications for particle physics, cosmology etc.

The stress would be more on the physical aspects with a discussion of measurable effects in relation to other areas of physics (with a discussion of orders of magnitude and number involved). We would thus have for instance a discussion of the analogy between torsion and magnetism (with consequences for astrophysics and cosmology), we will consider the Dirac equation in general relativity with torsion developed in a gauge theoretic manner with its implications for weak interactions and for strong interactions. We could have clear cut alternative ways of unifying gravity with electroweak and strong interactions by an energy dependent spin torsion coupling constant. The idea that all interactions can be understood as originating in spin curvature coupling is discussed.

The Maxwell equations when coupled to gravity in a space-time with torsion also gives rise to novel effects, there being effects due to polarisation of photons and neutrinos. We have analogy of Faraday polarisation, and other gravimagnetic effects distinct from Lense-Thirring precession, wmith is often confused which such effects.

Non-singular solutions of electrodynamics with gravity and torsion give rise to the possibility of finite classical stable models of elementary particles like the electron, the charge being derived from spin and magnetic moment.

We have a discussion of the possibility of cosmological models with torsion providing a possible solution of the cosmological constant problem.

Finally it is shown that torsion can lead directly to the quantization of space-time itself, the defects in space-time induced by torsion behaving like a string with a tension determined by the spin-torsion coupling in space-time.

An energy dependent tension at various energy scales, then gives the fundamental couplimg constants of the various interactions. This direct quantization of space-time leads not only to a minimal length but also to a minimal time with interesting consequences for early universe cosmology and physics of evaporating bleck holes. The quantum general relativistic principles take on an especially transparent and consistent form when torsion is taken into consideration. Thus the inclusion of torsion effects on space-time (after all this happens when intrinsic spin, a universal property of elementary particle like mass, is also a source of gravity) could help to resolve some of the complex conceptual issues inherent in the unification of quantum mechanics with general relativity. In addition it shows promise of resolving some fundamental problems in particle physics and cosmology. Some connecting links with twistors are also considered in this context.

It is hoped that this book, while presenting all the above aspects and more in concise terms with emphasis on the physical principles involved, would help to bring these interesting topics within range of physicists working on diverse fields like gravitation, cosmology and elementary particles.

We strongly believe that a book of this sort would draw many more physicists from different areas into this field.

Venzo de Sabbata
C.Sivaram

CONTENTS

CHAPTER I

SPIN AND TORSION

1. Motivations

We like to speak of torsion because we think that it is a necessary generalization of the Einstein theory of relativity that is a generalization of the Riemannian geometry of the space-time. This generalization goes back to a slight modification of the Einstein theory proposed in the 1922-23 by Cartan [1 - 4]. According to Trautman "the Einstein-Cartan theory is the simplest and the most natural modification of the original Einstein theory of gravitation" [5]. But why should one try to propose alternative or more general gravitation theories when, as we know, the general theory of relativity is the simplest gravitation theory which is in agreement with all experimental facts in the domain of macrophysics including the more recent experiment on time-delay, with radar on Mercury and Venus and other sophysticated experiments in solar system?

The motivations are mainly of theoretical character: in fact we are, at first sight, in the presence of two very different worlds: on the one hand we have at a microscopical level the strong and weak interactions while the gravitational interaction is the weakest and seems not to play any role at that level; on the other hand all known interactions, except gravitation, that is strong, weak and electromagnetic are well described within the framework of relativistic quantum field theory in flat Minkowski space-time. So at first sight it seems that gravitation has no effects when we are concerned with elementary particle physics. But today we know that this is not true: in fact if we consider the quantum theory in curved space-time instead of in flat Minkowski space-time, we have some very important new effects (neutron interferometry, spin-spin contact interactions, non-linear spinor effects etc.) and, moreover, when we go at a microphysical level, that is when we are concerned with elementary particle physics, we realize that the role of gravitation becomes very important and necessary for instance when we have to do with the early universe.

In the very early universe the cosmological problem is strictly connected with elementary particle physics. But then we must pay attention to this question: when we consider together general relativity and elementary particle physics, these latter described by quantum field theory, we are obliged to take into account not only the mass of elementary particles, but also the spin. In fact the elementary particles are characterized not only by mass but also by the spin which occurs in units of $\hbar/2$. Mass and spin are two elementary and independent original concepts: as a mass distribution in a space-time is described by energy-momentum tensor, so a spin distribution is described in a field theory by a spin density tensor. As the mass is connected with the curvature of space-time, the spin will be connected with another geometrical property of the space-time so that we must modify consequently the general theory of relativity in order to connect this new geometrical property with the spin density tensor. In this way we are led to the notion of torsion. In fact all elementary particles can be classified by means of the irreducible unitary representation of the Poincaré group and can be labelled with translational part of the Poincaré group, while spin is connected with the rotational part. In a classical field theory, mass corresponds to a canonical stress-energy-momentum tensor, and spin to a canonical spin tensor. The dynamical relation between the stress-energy-momentum tensor and curvature is expressed in general relativity by Einstein equations; one feels here a need for an analogous dynamical relation including spin density tensor. So as this is impossible in the framework of the general relativity we are forced to introduce this new geometrical property that we call torsion. We can say that as the mass is responsible for curvature , spin is responsible for torsion. We now will see from a formal point of view in which way we must modify the general relativity theory and how it represents a slight modification of this theory: in fact the main point is to assume an affine connection asymmetric instead of the symmetric connection that we have in Einstein theory (the Christoffel symbols). Torsion is in fact connected with the antisymmetric part of the affine connection as we shall see.

To sum up: when we are dealing with a microphysical realm we find that, besides the mass, the spin comes into play and then has to be considered as the source of a gravitational field (as the mass): remember that when we speak of a gravitational field, we means a field inseparably coupled to the geometry that is we speak of the structure of the space-time. In analogy to the mass where the energy-momentum tensor is coupled to the metric, we will expect that the spin-density tensor is coupled to some geometrical quantity of the space-time (a quantity that should relate to the rotational degrees of freedom in space-time).

In this way we are led to a generalization of a Riemann space-time. We will see very briefly which are the new geometrical properties of the torsion as all the development will follow very nearby the structure of general relativity. We may also notice that Cartan propose to relate the torsion tensor to the density of intrinsic angular momentum since 1922 well before the introdction of the modern concept of spin by Uhlenbeck and Goudsmit [6]. This fact may have been the reason for which the Cartan hypothesis has been ignored for so long a time, and then, when it has been considered, was connected not with the spin but with electromagnetic field in the attempt to unify the gravitational and electromagnetic interaction. At last it should be noted that with the introduction of torsion, with this small departure from Einstein's general relativity, the fields equations in empty space are the same so that the majority of its experimental verifiable consequences for the solar system cannot be distinguished from the prediction of general relativity. On the other hand we will see that we will have many important new consequences: for instance when we apply this Einstein-Cartan theory in the ambit of the cosmology, especially in the early universe; or when considering superdense objects as neutron stars or black holes where we have strong internal magnetic fields which may give rise to alignment of spins or viceversa we will see that torsion interaction may align spins giving rise to magnetic field. So we will consider the problem of magnetic field in the universe, the neutrino magnetic moment, the connection of torsion with quantum effects, the problem of cosmological constant, to quote only some of the problem where torsion appear to be important. But let us first of all introduce torsion, that is introduce spin in general relativity as a dynamical quantity.

2. Riemann-Cartan geometry

Torsion as the antisymmetric part of an asymmetric affine connection was introduced by Cartan, as we have said [1 - 4]. The definition of Cartan torsion is then the antisymmetric part of the affine connection $\Gamma_{\alpha\beta}{}^{\mu}$ that is

$$Q_{\alpha\beta.}{}^{\mu} = (1/2)(\Gamma^{\mu}_{\alpha\beta} - \Gamma^{\mu}_{\alpha\beta}) \equiv \Gamma^{\mu}{}_{[\alpha\beta]} \qquad (1.1)$$

and has a tensor character: Cartan had some idea then it might be connected with the intrinsic angular momentum of matter but only later it becomes clear that the energy-momentum tensor of massive spinning fields as for instance Dirac field, must be asymmetric (see Papapetrou, [7,8]). Now we know that the Einstein field equations show how the stress energy-momentum tensor of matter generates a curvature of space-time. But is only this tensor the source of geometry? or is there some other dynamical quantities which influence the geometry of the space-time?

We know that the tetrad field is determined in each point of the space-time by metric tensor up to the six-parameter group of rotations and it is possible to treat gravitation as a compensating field of the group of general coordinate transformation and tetrad rotation (in a similar way to the treatment of the electromagnetic field as a compensating field of the group of gauge transformations of charged fields [9], [10]). In this way one can introduce the spin as a dynamical quantity in the theory of gravitation and one is led to an extension of the geometric structure of the space-time. The Cartan idea about a relation between torsion and spin may be supported considering that in special relativity theory we have to do with the Poincaré group. The Poincaré group is a product of the Lorentz group and the group of translations. Now the mass is connected with the translational part of the Poincaré group while the spin is connected with the rotational part. In classical field theory mass corresponds to a canonical stress-energy-momentum tensor and spin to a canonical spin tensor. In general relativity the dynamical relation between the stress-energy-momentum tensor and curvature is expressed by the Einstein equations so one may think that it must be an analogous dynamical relation including spin tensor, a relation that would connect the spin tensor with another geometric property of the space-time. This obviously is not possible in the ambit of general relativity so that one is necessitated to modify the theory in order to introduce torsion and relate it to the spin tensor (see for instance Hehl et al.[11]) .

When this is done one arrives at an interesting link between the theory of gravitation and special relativity: in fact one can show that torsion is related to translations in the tangent space of the manifold in a similar fashion as the curvature is related to Lorentz transformation. If $\mathfrak{L}$ denotes the material Lagrangian density in general relativity we have for the dynamical stress-energy-momentum tensor:

$$T^{\mu\nu} = \frac{2}{\sqrt{-g}} \frac{\delta \mathfrak{L}}{\delta g_{\mu\nu}} \qquad (1.2)$$

($\sqrt{-g} = (-\det g_{\mu\nu})^{1/2}$ is the root of the determinant of the metric).

Now if $\mathfrak{L}$ depends not only on the tensor $g_{\mu\nu}$ and its derivatives but also on a new geometrical quantity $K_{\mu\nu}{}^{\alpha}$, we have the following dynamical definition of spin:

$$S_{\alpha}^{\cdot\mu\nu} = \frac{1}{\sqrt{-g}} \frac{\delta \mathfrak{L}}{\delta K_{\mu\nu}{}^{\alpha}} \qquad (1.3)$$

$K_{\mu\nu.}{}^{\alpha}$ is called the contorsion tensor and, as we shall see later on, is connected with torsion tensor defined in (1.1) by:

$$K_{\mu\nu.}{}^{\alpha} = -Q_{\mu\nu.}{}^{\alpha} - Q^{\alpha}{}_{.\nu\mu} + Q_{\mu.\nu}{}^{\alpha} = -K_{\nu.\mu}{}^{\alpha} \qquad (1.4)$$

so that we can write for the affine connection

$$\Gamma_{\mu\nu}{}^{\alpha} = \left\{ {\alpha \atop \mu\nu} \right\} - K_{\mu\nu.}{}^{\alpha} \qquad (1.5)$$

It is usual to denote the Riemann-Cartan space-time of the Einstein-Cartan theory (when an affine asymmetric connection is introduced) as U_4 to distinguish it from the Riemann space-time (with a symmetric affine connection) which is denoted by V_4. It is then possible to go from V_4 to U_4 by simply substituting everywhere the asymmetric affine connection in the place of Christoffel symbols. When $K_{\mu\nu.}{}^{\alpha} = 0$ we are back in the Riemannian space V_4.

Consider now in U_4 a differentiable manifold in 4-dimensions in which is defined a metric $g_{\mu\nu}(x)$ which provides the distance ds between two nearby points x^μ and $x^\mu + dx^\mu$

$$ds^2 = g_{\mu\nu}(x)dx^\mu dx^\nu \qquad (1.6)$$

and in which is defined also an affine connection $\Gamma^{\mu}_{\alpha\beta}$ which provides the infinitesimal variation of a vector B^μ during the parallel transport from x^μ to $x^\mu + dx^\mu$:

$$\delta B^{\mu} = - \Gamma^{\mu}_{\alpha\beta}(x) B^{\beta} dx^{\alpha} \tag{1.7}$$

Now if B^{μ} are the components of the vector in x^{μ} and $B^{\mu} + dB^{\mu}$ the components in $x^{\mu} + dx^{\mu}$ we have

$$dB^{\mu} - \delta B^{\mu} = (\partial B^{\mu}/\partial x^{\alpha}) dx^{\alpha} + \Gamma^{\mu}_{\alpha\beta} B^{\mu} dx^{\alpha} \tag{1.8}$$

and is easy to see that the antisymmetric part of the affine connection , that is $\Gamma^{\mu}_{[\alpha\beta]}$ is a tensor.

In order to explain the geometric meaning of the antisymmetric part of the connection, we can say that from the geometrical point of view the basic property of the torsion is that the closed paths are open in the tangent space while, as we know, in the case of the curvature a vector which goes along a closed path changes direction when returns to the starting point (Fig.1a); in the case of torsion the vector is suffering a translation (Fig.1b); obviously if both curvature and torsion are present, the vector will have a rotation and a translation (fig.1c).

This leads to important physical consequences. That is the modification of the starting point is minimal (a asymmetric connection instead of the symmetric one of the Einstein theory) but the physical consequences are significant.

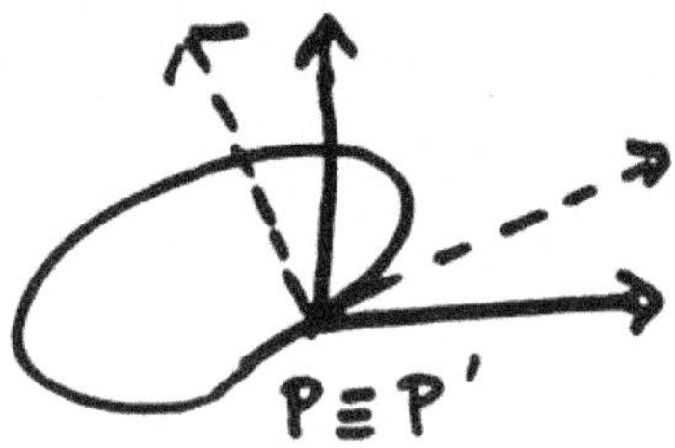

Fig.1a - curvature

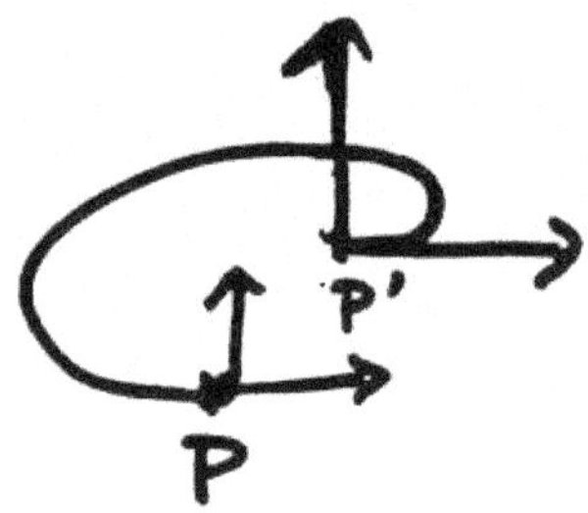

Fig.1b - torsion

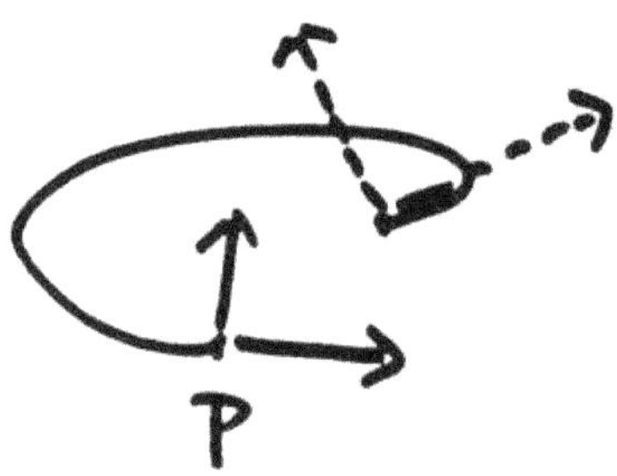

Fig.1c - curvature and torsion

Fig.1a,b,c, - when a vector describes an infinitesimal closed path, and when this path is developed in the flat space tangent to the manifold, we have a rotation (if there is only curvature), or a translation (if there is only torsion) or both (if there is curvature and torsion)

We will briefly illustrate what we have said: i.e. we will show that if we construct infinitesimal parallelograms in the tangent space at a point P, in general they do not close (Fig.2), the failure of closing being proportional to torsion. In other words a closed contour (P, A_1 , A_2 , A_3) in an U_4 manifold becomes in general a non-closed contour (P, B_1 , B_2 , B_3) in the flat space E tangent to U_4 in P (Fig.2)

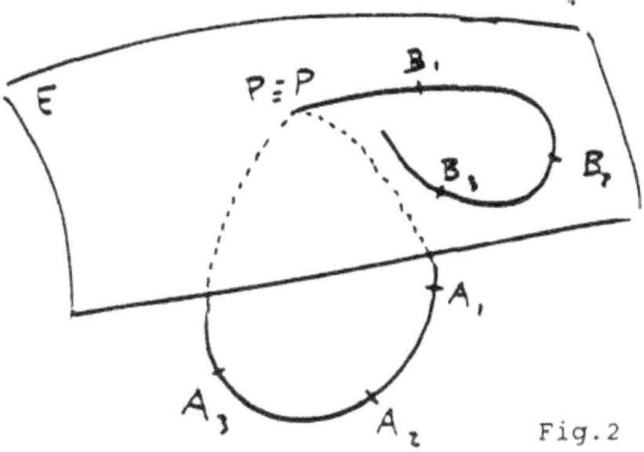

Fig.2

Consider in fact the infinitesimal parallelogram PACB in the tangent space E, and put $\overline{PA}$ = dx′, $\overline{PB}$ = dx′′ (Fig.3)

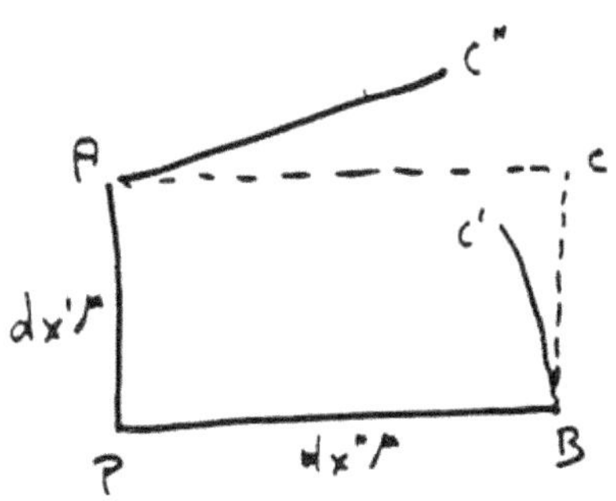

Fig.3

Displacing dx′ along dx′′ in the corresponding U_4 manifold we have

$$\delta(dx'^{\mu}) = -\Gamma_{\alpha\beta}{}^{\mu}\, dx'^{\beta} dx''^{\alpha} \tag{1.9}$$

while for the displacement of dx′′ along dx′

$$\delta(dx''^{\mu}) = -\Gamma_{\alpha\beta}{}^{\mu} dx''^{\beta} dx'^{\alpha} \tag{1.10}$$

Therefore

$$\delta(dx''^{\mu}) - \delta(dx'^{\mu}) = -\Gamma_{\alpha\beta}{}^{\mu}(dx'^{\beta} dx''^{\alpha} - dx'^{\alpha} dx''^{\beta})$$

which we can write as

$$= -(\Gamma_{\alpha\beta}{}^{\mu} - \Gamma_{\beta\alpha}{}^{\mu}) dx'^{\beta} dx''^{\alpha} = 2\, Q_{\alpha\beta}{}^{\mu}\, dx'^{\beta} dx''^{\alpha} \tag{1.11}$$

and as in the presence of torsion $Q_{\alpha\beta}{}^{\mu} \neq 0$, we have in general that the points C′, C′′ and C do not coincide so that the parallelogram is not closed. In the Riemannian space the Γ are symmetric (the Christoffel symbols) and then $Q = 0$.

This property that we have in the Riemannian space is then clearly a consequence of the symmetry of the affine connection.. This is no more valid if the connection is asymmetric: in this last case we have two different points and, as we have seen, the parallelogram is not closed; this is just the geometric characteristic of torsion.

We will see that the closure failure behaves as dislocations (or defects) in crystals and then torsion plays the role of defect density and the defect in space-time topology leads naturally to a possible quantization of space-time. We think in fact that when one likes to quantize gravity, one cannot quantize gravity in analogy to the other interactions in nature because, according to Einstein, there is no gravity force but curvature of space-time; so if one like to quantize gravity one must go towards the quantization of space-time itself. The torsion seems to represent a way to go through the quantization of space-time.

But before that, some words on the mathematical side: in order to find the field equations one proceeds as in the Einstein case; the only difference is that the affine connection Γ is not symmetric.

First of all it is important to note that in U_4 the geodesic equation

$$d^2x^\mu/ds^2 + \left\{ {\mu \atop \alpha\beta} \right\} (dx^\alpha/ds)(dx^\beta/ds) = 0 \qquad (1.12)$$

can be obtained, as in V_4 , by the variational principle

$$\delta \int (g_{\mu\nu} dx^\mu dx^\nu)^{1/2} = 0 \qquad (1.13)$$

It defines curves of extremal length with respect to the metric, and it is different, in general, from the autoparallel equation

$$d^2x^\mu/ds^2 + \Gamma_{\alpha\beta}{}^\mu (dx^\alpha/ds)(dx^\beta/ds) = 0 \qquad (1.14)$$

which defines curves along which a vector is displaced

parallelely to itself, according to the affine connection of the manifold. In fact only the symmetric part of the affine connection enters in eq.(1.14) (because of the symmetry of the product $dx^\alpha dx^\beta = d\, x^\beta dx^\alpha$) and the symmetric part do not coincide with the Christoffel symbols; we have:

$$d^2x^\mu/ds^2 + \Gamma_{(\alpha\beta)}{}^\mu (dx^\alpha/ds)(dx^\beta/ds) =$$
$$= d^2x^\mu/ds^2 + \left[\left\{ {\mu \atop \alpha\beta} \right\} + 2\, Q^\mu{}_{(\alpha\beta)}\right](dx^\alpha/ds)(ds^\beta/ds) \qquad (1.15)$$

We will see that neither the geodesic equation (1.12) nor the autoparallel equation (1.14) represent the motion equation.

Continuing along these lines, one introduces the curvature tensor which is defined as in a Riemannian space, using however the U_4 connection instead of Christoffel symbols. Therefore:

$$R_{\alpha\beta\mu}{}^\nu = \partial_\alpha \Gamma_{\beta\mu}{}^\nu - \partial_\beta \Gamma_{\alpha\mu}{}^\nu + \Gamma_{\alpha\lambda}{}^\nu \Gamma_{\beta\mu}{}^\lambda - \Gamma_{\beta\lambda}{}^\nu \Gamma_{\alpha\mu}{}^\lambda \qquad (1.16)$$

and using this explicit expression one may verify that the following identities are satisfied:

$$R_{(\alpha\beta)\mu\nu} = 0 = R_{\alpha\beta(\mu\nu)} \qquad (1.17)$$
$$R_{[\alpha\beta\mu]}{}^\nu = 2\, \nabla_{[\alpha} Q_{\beta\mu]}{}^\nu - 4\, Q_{[\alpha\beta}{}^\lambda\, Q_{\mu]\lambda}{}^\nu \qquad (1.18)$$

The Bianchi identities are generalized as follows:

$$\nabla_{[\alpha} R_{\beta\mu]\nu}{}^\lambda = 2\, Q_{[\alpha\beta}{}^\rho R_{\mu]\rho\nu}{}^\lambda \qquad (1.19)$$

and the Einstein tensor is defined, as usual:

$$G_{\mu\nu} = R_{\mu\nu} - (1/2)g_{\mu\nu}R \tag{1.20}$$

where $R_{\mu\nu} = R_{\alpha\mu\nu}{}^{\alpha}$ is the Ricci tensor and $R = R_{\mu}{}^{\mu}$. So formally the Einstein tensor is the same as in Riemann space-time but now is important to stress that, in the Riemann-Cartan space-time $G_{\mu\nu}$ and $R_{\mu\nu}$ are no longer symmetric tensors. This new geometrical property of the Einstein tensor is related to the conservation of the total angular momentum in the presence of a non zero intrinsic spin density. At this point, starting from a Lagrangian which will be not only function of the metric tensor and its derivatives but also of the torsion tensor (and, if that is the case, that is if we consider a propagating torsion, also of its derivatives) and applying the usual variational method, we arrive at the field equations:

$$\begin{aligned} G^{\mu\nu} &= \chi\, t^{\mu\nu} \\ T^{\alpha\mu\nu} &= \chi\, S^{\alpha\mu\nu} \end{aligned} \tag{1.21}$$

where the first equation is similar to Einstein equation but $t^{\mu\nu}$ contains also the spin density tensor, while the second one is the new relation which connects the geometrical quantity $T^{\alpha\mu\nu}$ (defined as $T_{\alpha}{}^{\mu\nu} = - (1/2)(-g)^{1/2}\delta R/\delta K_{\nu\mu}{}^{\alpha}$ being R the scalar curvature and K the contorsion tensor given in (1.4)) with the spin density tensor as defined in (1.3).

The equation of motion for the Einstein-Cartan theory is

$$(dp^{\mu}/ds) + \left\{ {\mu \atop \nu\alpha} \right\} p^{\alpha}u^{\nu} + F^{\mu} = 0 \qquad (1.22)$$

where

$$F^{\mu} = (dt/ds)\int d^3x'(-g)^{1/2}\phi^{\mu} \qquad (1.23)$$

represents a force term arising from the interaction of the spin with the geometry of the U_4 manifold.

If the test particle is spinless (or if the macroscopic test body is unpolarized) then $S_{\mu\nu\alpha} = 0$, $\phi^{\mu} = 0$, $t^{\mu\nu} = T^{\mu\nu}$ and the equation of motion reduces to the usual geodesic equation

$$(du^{\mu}/ds) + \left\{ {\mu \atop \nu\alpha} \right\} u^{\nu}u^{\alpha} = 0 \qquad (1.24)$$

References

[1] E.Cartan - Compt.Rend.174, 437, 593 (1922)
[2] E.Cartan - Ann.Ec.Norm.40, 325 (1923)
[3] E.Caratn - Ann.Ec.Norm.41, 1 (1924)
[4] E.Cartan - "Sur les variables à connexion affine et la théorie de la relativité généralisée" ed.Gauthier Villard 1955
[5] A.Trautman - "Theory of Gravitation", preprint IFT/72/25, Warsaw Uni.: read at the Symposium "On the Development of the Physicist's Conception of Nature', Miramare, Trieste 1972
[6] G.H.Uhlenbeck and S.A.Goudsmit - Physica 5, 266 (1925)
[7] A.Papapetrou - Philos.Mag.40, 937 (1949)

[8] A.Papapetrou - Proc.Roy.Soc.(London) A209, 249 (1951)
[9] D.W.Sciama - in "Recent Developments in General Relativity", Pergamon Press and PWN, Warsaw 1962, pag.415
[10] T.W.Kibble - J.Math.Phys.2, 212 (1961)
[11] F.W.Hehl, P.von der Heyde, G.D.Kerlick and J.M.Nester Rev.Mod.Phys.48, 393 (1976)

CHAPTER II

TORSION AND MAGNETISM

1. Celestial magnetic field

We like to consider two astrophysical empirical laws, and try to give some theoretical basis for them and to show that the fine structure constant α is a typical dimensionless parameter not only of atomic physics but also of gravitational physics.

The astrophysical laws for which we are concerned are:
a) the relation between intrinsic angular momentum S and the magnetic moment U of stars [1],[2],[3] that may be written as

$$S = qU \tag{2.1}$$

where q is a universal constant;
b) the relation between total angular momentum J and the mass M of astronomical system [4] that can be written in the form

$$J = pM^2 \tag{2.2}$$

where p is a universal constant.

From the experimental data of ref.[3] which report the maximum and minimum values of the measured magnetic fields we can obtain for q the mean value $q = 1.3\cdot 10^{15}\ g^{1/2}cm^{-1/2}$ while the data of ref.[5] gives for p the value $p = 7.9\cdot 10^{-16}\ g^{-1}cm^{2}sec^{-1}$. The remarkable result is that both relations seem to indicate that the fine structure constant α is the typical dimensionless parameter not only of atomic physics but also of gravitational physics.

As is well known, Blackett [1] on the basis of some observed ratio of the magnetic-dipole moment to the angular momentum of some planets and stars, established the empirical relation

$$U = \beta \, (\sqrt{G}/c) S \tag{2.3}$$

where β is a numerical coefficient of the order of unity.

The other empirical relation (2.2) which relates the total angular momentum J and the mass M of the astronomical systems was extensively investigated by Brosche [4] and by Wesson [5].

At first sight these two empirical laws seem to be completely unrelated but we will show that a close connection can be established between these two laws and since the relation (1.1) can be given a theoretical basis with the introduction of the spin-torsion interaction and some considerations concerning the early universe, also relation (2.2) acquires, indirectly, a cosmological interpretation. First of all, in view of a possible unification of all the fundamental interactions, Wesson [5] argued that a new dimensional constant is to be introduced in gravitational physics, so that its combination with the Newton coupling constant G and the light velocity c can form a dimensionless parameter characteristic of the gravitational interaction.

The new universal constant, according to Wesson, is p defined by the mass-angular momentum relation (2.2), and the dimensionless number we can obtain is

$$\beta = G/cp \qquad (2.4)$$

Wesson assumedthat β is typical parameter of gravitational physics, like the fine structure constant $\alpha = e^2/\hbar c$ is typical of microphysics. Now here we wish to point out that also the empirical law (2.1) provides a new universal constant q and starting with q we can form another dimensionless number [6]:

$$\gamma = c^2/q^2G \qquad (2.5)$$

which we could choose, instead of β, as the dimensionless constant characteristic of the gravitational interaction. The remarkable result we wish to stress is that, from experimental values of p and q one gets

$$\beta \simeq 0.28\cdot10^{-2} \quad , \qquad \gamma \simeq 0.79\cdot10^{-2} \qquad (2.6)$$

and then $\beta \simeq \gamma$ up to an order of magnitude.

If one does not believe that this numerical coincidence is accidental, then this result seems to suggest that the two set of constants (G,c,p) and (G,c,q) are equivalent in the sense that both define the same dimensionless parameter $\beta = \gamma$ characteristic of gravitation.

It follows that the two constants p and q are not independent since equating the two definitions (2.4) and (2.5) we obtain the relation

$$q^2/p = c^3/G^2 \qquad (2.7)$$

which establishes a connection between the two empirical laws, and suggest the possibility they have a common origin.

Therefore once that a theoretical explanation is provided for one of these two laws, and one of the two constants p or q is fixed, then from (2.7) another fundamental constant can be deduced or, equivalently, the existence of another law of nature can be predicted.

Now, as previously said, we have in fact a cosmological explanation of the first law (2.1), the Blackett law, if we introduce the spin-torsion interaction during the early stages of our universe. The Blackett law appear as a consequence of a resonant spin alignement during the early stage of the universe.

We do not go in the detail of calculation that one can find in the ref.[7],[8], and will give only an outline.

Before to do that we can observe that Blackett was led to the relation S = qU observing that we know a fundamental relation between magnetic moment and spin angular momentum for microscopic charged particle: the Bohr magneton e/2mc; this is enormously greater respect to ~ 10^{-15} given for macroscopic bodies. This fact is due to the enormous charge density of electron. If on the contrary we substitute the gravitational mass $G^{1/2}m$ at the place of the electric charge, where is just the gravitational mass $G^{1/2}m$ responsible for producing magnetic field of a rotating body, then we would expect the ratio of the magnetic moment to the angular momentum to be smaller then the Bohr magneton by the ratio $G^{1/2}m/e$ and in this way we are led to the value of $\sim 10^{-15}cm^{1/2}g^{-1/2}$, which is the value seen above for many celestial bodies.

Now we will see that, as said, the relation (2.1) has a cosmological significance since it can be deduced in a very simple way as a natural consequence of the initial conditions of the universe provided that torsion is introduced into the gravitational equations according to the Einstein-Cartan theory.

2. Spin-Torsion Lagrangian

It is in fact well known that the spin-torsion interaction Lagrangian in a Riemann-Cartan space-time is

$$\mathfrak{L} = -(1/2)\ S^{kij}\ K_{ijk} \qquad (2.8)$$

where S^{kij} is the canonical spin tensor and K_{ijk} is the contorsion tensor defined in (1.5).

Eq.(2.8) may be rewritten as (since $S_{ijk} = -S_{jik}$)

$$\mathfrak{L} = (1/2)\ S^{kij}[Q_{kij} - 2Q_{j(ki)}] = (1/2)S^{kij}Q_{kij} \qquad (2.9)$$

(Remember that torsion is related to the spin density tensor $J_{ij}{}^{k}$ by the Einstein-Cartan field equation

$$Q_{ij}{}^{k} = (8\pi G/c^3)[J_{ij}{}^{k} - (1/2)\delta^k_i J_{lj}{}^{l} - (1/2)\delta^k_j J_{il}{}^{l}] \quad (2.10)$$

and that $\quad S_{ij}{}^{k} = \int d^3x \sqrt{-g}\ J_{ij}{}^{k} \qquad$).

In the case of Dirac particle (the spin is a totally antisymmetric tensor) we have

$$S^{ijk} = \eta^{ijkl} S_l$$

(where η^{ijkl} is the completely antisymmetric symbol) , and the interaction Lagrangian contains only the axial-vector part of the torsion tensor

$$\mathcal{L} = S_k Q^k \qquad (2.11)$$

(being $Q_k = (1/2)\eta_{kijl}Q^{ijl}$) and going to the particle rest system, where $S^k = (0,\bar{S})$, the interaction energy between a spin $\bar{S}$ and a torsion $\bar{Q}$ may be written as

$$E = -\mathcal{L} = -\bar{S}\cdot\bar{Q} \qquad (2.12)$$

We can observe that this formula is formally analogous to the interaction energy of a magnetic moment $\bar{\mu}$ in a constant magnetic field $\bar{H}$ namely:

$$E = -\bar{\mu}\cdot\bar{H} \qquad (2.13)$$

so that we can understand, without going through the calculation(see ref.[7] and [8]) that torsion induces spin alignement as well as a magnetic field and if to each spin is associated a magnetic moment, torsion can also produce a magnetic field inside the matter. Applying these consideration to the early universe we find that to each portion V of hadronic matter (V is the volume of the portion) containing a total angular momentum density J = S/V and then a torsion Q = $4\pi G_f J/c^2$, it was associated a magnetic field (produced by the nucleons orientation)

$$B = 8\pi/3c\ (2\ \alpha\ G_f)^{1/2} J \qquad (2.14)$$

where α is the fine structure constant and G_f the strong gravity constant (for the necessity of the introduction of strong gravity constant see [9] and [10]), so that we have a dipole magnetic moment

$$U \sim (\alpha\ G_f)^{1/2}\ S/c \qquad (2.15)$$

At this point we can observe that the same relation between angular and magnetic moment holds also in the case of a particle. In fact:

$$\mu/\hbar \simeq e/mc \simeq (\alpha\ G_f)^{1/2}/c \qquad (2.16)$$

This result fits nicely with Dirac's two metric hypothesis [12] that is the atomic metric operative at a microscopic level and the usual macroscopic metric connected with gravitation. These two metrics coincide at the beginning of the hadron era, but the atomic metric has not changed in time, while the macroscopic metric developed with the universe.

Therefore the ratio U/S for a particle was frozen to the

value (2.16) it had at that epoch ($t_0 = 10^{-23}$sec when, according to Dirac large number hypothesis [13] for which the gravitational constant decreases in time as $G \propto t^{-1}$, the gravitational constant coincides with the strong gravity constant G_f), while the magnetic field of a celestial body has developed in time up to its present value

$$U \sim (\alpha G)^{1/2} S/c \qquad (2.17)$$

This explain also why Blackett's experiment [14], which measured the magnetic field of a rapidly rotating body, provided a null result.

So we find the Blackett relation

$$S = qU \qquad (2.1)$$

with

$$q = c/(\alpha G)^{1/2} \qquad (2.18)$$

which is in remarkable agreement with astrophysical data (that is with $q = 1.3 \cdot 10^{15} g^{1/2} cm^{-1/2}$). Using now (2.18) and the relation (2.7)

$$q^2/p = c^3/G^2 \qquad (2.7)$$

connecting the two astrophysical laws, we can deduce that the ratio of the angular momentum over squared mass ($J = pm^2$) is a universal constant p given by

$$p \sim G/\alpha c \sim 3 \cdot 10^{-16} g^{-1} cm^2 sec^{-1} \qquad (2.19)$$

again in agreement with experimental data.

The remarkable result is that inserting into the definition (2.4) and (2.5) of the dimensionless numbers β and γ the values of p and q both obtained from our theoretical predictions (2.18) and (2.19), we obtain $\beta \sim \gamma \sim \alpha$ where $\alpha \sim 0.73 \cdot 10^{-2}$ is the fine structure constant.

This means that the two empirical, and seemingly unrelated, astrophysical laws (2.1) and (2.2) not only define the same constant $\beta = \gamma$ but also suggest that the dimensionless parameters of gravitation and microphysics are the same $\beta = \gamma = \alpha$. Therefore, according to Wesson arguments [5], they may represent a promising clue to the unification of gravitation with the other fundamental interaction: the dimensionless coupling constant for particle physics is the same as the dimensionless coupling constant for gravitation.

On these lines we can provide also a primordial magnetic field in the Universe. In fact if the universe is rotating and posesses a torsion field and a non vanishing total angular momentum density [15], then it must have a universal magnetic field B_U (produced by the orientation of the nucleons by the torsion-spin interaction):

$$B_U = (4/c)(2\alpha G)^{1/2} S_U/R_U^3 \qquad (2.20)$$

where, if we refer to value of today, we have $R_U \sim 10^{28}$cm (the Hubble radius) and $S_U \sim 10^{120}\hbar$, the total angular momentum of the Universe, being related to the spin of elementary particle by the condition (see [9]) $S_U = \hbar\ (R_U/r)^3$ (being r the Compton wave-length of a hadron $r = \hbar/mc$) so that we find

$$B_U \sim 10^{-6}\text{Gauss} \qquad (2.21)$$

At this point we like to observe that it is remarkable that the energy density of the interstellar magnetic field $B^2_{int}/8\pi$ is comparable to that of microwave background radiation i.e. $B^2_{int}/8\pi = aT^4$ (where a is the Stefan-Boltzmann constant) so that we have

$$B_{int} \sim (8\pi aT^4)^{1/2} = (8\pi a)^{1/2}T^2 \qquad (2.22)$$

which for $T \simeq 3$ K gives $B_{int} \approx 3\cdot10^{-6}$G.

This suggest some kind of equipartition between cosmic magnetic field and radiation field energies. It also strongly suggests that analogously to the radiation background we have the existence of a primordial magnetic field at the time of big bang providing a natural fossil origin for galactic and intergalactic magnetic fields. Such a field may affect the formation of galaxies and the nucleosynthesis of primordial elements.

This equipartition between magnetic and thermal radiation background energies observed at the present epoch has suggested to many authors that such equipartition was also prevalent at the earliest epochs of the big bang. In particular at the earliest Planck epoch $t_{Pl} = (\hbar G/c^5)^{1/2} \sim 10^{-43}$s, where the energy density was $\varepsilon_{Pl} = c^7/\hbar G^2 \sim 10^{115}$erg cm^{-3}, the primeval magnetic field would have had the highest value

$$B_{Pl} \approx (8\pi c^7/\hbar G^2)^{1/2} \approx 10^{58}G \qquad (2.23)$$

It is usually assumed that the magnetic flux $\phi = B\ R^2$ is conserved during adiabatic expansion of the universe, so that the dependence of the magnetic field on the expansion time t and the background temperature T has the form

$$B\ t = B\ T^{-2} \qquad (2.24)$$

Equation (2.24) would then imply that in the case of flux conservation, B at the nucleosynthetic epoch (B_{nucl}) of t ≈ 1s, would have the value

$$B_{nucl} \approx 10^{15}G \qquad (2.25)$$

However a field of the value given by (2.25) at the time of nucleosynthesis would have drastic consequences for the abundances of the light elements as the magnetic field contribute an additional energy density source for increasing the expansion rate of the universe. The maximal allowed magnetic field can be determined equating the weak interaction equilibrium rate among the nuclei and leptons to the expansion rate caused by the field. Thus this gives [16] for KT = 1 MeV at the nucleosynthetic era

$$B_{nucl} \leq 3\cdot 10^{12}G \qquad (2.26)$$

A more stringent limit can be obtained by noting that even an extra contribution of about 5% to the overall energy density of the radiation would alter the expansion rate significantly to change the helium and other abundance and would imply $B_{nucl} \leq 10^{12}G$. So (2.25) would violate this bound indicating that the assumption of conserved magnetic flux during the expansion may not be valid.

Many mechanism have been proposed to generate the correct primeval field but all proposed mechanism do not give the field we expect. For instance in the context of inflationary models wherein in the post-Planck epoch the universe undergoes an exponential expansion phase by a factor of $\approx$ $\exp(H_{in}t)$ ($H_{in}t \approx 65$ for the inflation to be sufficient to solve the flatness and horizon problems [17]) this would imply that the magnetic field of $B_{Pl} \approx 10^{58}$Gauss would decay as $B_{Pl}\exp(-2H_{in}t)$ in the inflation era (H_{in} is constant in the inflation epoch, so that the field decays exponentially).

Thus it is implied that B_{Pl} would reduce by at least a huge factor of $e^{130} \approx 10^{56}$ leading to a maximum magnetic field of only $\approx 10^{2}$Gauss at the end of inflation and since during the radiation epoch the universe expands by a factor of about 10^{20}, this would reduce the present value of the cosmic field to only 10^{-38}Gauss, which is far too a low value. Other models do not work better.

But returning to our previous considerations, i.e. considering torsion-spin interaction, we have found for the magnetic field in terms of the spin density σ:

$$B \approx (8\pi/3c)(2\alpha G)^{1/2}\,\sigma \qquad (2.27)$$

By the hypothesis of tensor-meson dominance, hot hadronic matter undergoes strong gravitational interactions [9],[18] (by exchange of f-mesons) with a strong coupling constant $G_f \approx \hbar c/m^2 \approx 10^{39}G$ and so G must be replaced by G_f in eq.(2.27).

Thus for a neutron star with all the nucleon spins aligned by the strong gravity spin-torsion interaction, eq.(2.27) would give a magnetic field $B \approx 10^{15}$Gauss [19]. This is equivalent to the alignement of the magnetic moments $\mu_n \approx e\hbar/2mc \approx 10^{-23}$ erg/Gauss of all the nucleons in the neutron star. Thus the spin-torsion interaction also naturally accounts for the Blackett's empirical relation eq.(2.1).

At the stage of the hadron era (i.e. prior to the nucleosyntehesis period), the universe had a density of the same order of magnitude as the nuclear density and again one considers the strong gravitational torsion interaction between the hadrons. The radius of the universe at that stage (see for instance [9]) was $\approx 10^{13}$cm, so that if all the magnetic moments of the 10^{78} nucleons are aligned, this would give a magnetic field of $B_h \approx 10^{19}$Gauss. Essentially this is equivalent to having G_f instead of G in eq. (2.23). So when the universe was at a temperature of $T_h \approx 10^{13}$ K in the hadron era, the magnetic field induced by the alignement of the nuclear spins by the strong gravity spin-torsion interaction was $B_h \approx 10^{19}$Gauss. At the nucleosynthetic stage when $T_n \approx$

10^9 K, then eq.(2.24) would imply that the magnetic field at the time when the light elements were being synthesized in the big bang, was $B_{nucl} \leq 10^{11}$Gauss, which is consistent with the constraints imposed by eq.(2.26). The present value B_0 of the field, i.e. the magnetic field at the present epoch ($T_o \approx$ 3 K) would then be (from eq.(2.24))

$$B_o \approx B_{nucl}(T_o/T_n)^2 \approx 10^{-7}\text{Gauss} \qquad (2.28)$$

which is a reasonable value as compared to many of the mechanisms discussed earlier.

The presence of such a primeval magnetic field has interesting implications for galaxy formation. The critical Jean's length and mass are modified with the magnetic field to

$$\begin{aligned} l_{J,B} &\approx v_A/(G\,\rho)^{1/2} \quad \text{and} \\ M_{J,B} &\approx (\pi/6)B^3\rho^{-2}G^{-3/2} \end{aligned} \qquad (2.29)$$

where v_A is the Alfven speed. The additional pressure and density introduced by the magnetic field increases the resultant critical Jeans mass and its inclusion could help to account for observed physical parameters of galaxies. Estimate for $M_{J,B}$ from eq.(2.29) are in rough agreement with mass estimate for dwarf galaxies which consequently formed first, the bigger structure forming at later epoch.

The existence of the all pervasive and isotropic microwave cosmic background radiation of ≈ 3 K provides strong evidence of a dense hot early phase for the universe. It is striking that apart from the onnipresent thermal radiation background we have the ubiquitous existence of magnetic fields observed to be present in interstellar space with an estimated field strength $B = 10^{-6}$Gauss and inferred to be responsible for a variety of phenomena (like polarization of starlight from alignement of dust grains, galactic radio emission, etc.) producing observable electromagnetic effects.

There is also a variety of evidence such as from the Faraday rotation of extragalactic radiosources that intergalagtic magnetic field about two orders lower in magnitude may be present. For detailed review of evidence see ref.[20]. Intergalactic magnetic fields in excess of 10^{-6} Gauss would cause noticeable anisotropies in the microwave background.

3. Torsion from magnetism

It is generally thought that magnetic fields which may be present in gravitationally collapsing objects or systems do not significantly affect the process of continued contraction for a sufficiently large mass. A basic reason for this supposition is that flux conservation implies that the magnetic field increases during contraction only as the two-thirds power of the matter density i.e. $B \propto \rho^{2/3}$ (area being proportional to $\rho^{-2/3}$) so that the magnetic pressure increases at most as $\rho^{4/3}$.

Since the pressure due to gravitational contraction goes as ρ^2, one sees that for sufficiently large ρ, magnetic fields would be inadequate to prevent the collapse to a singular state.

However this argument neglects the gravitational effects of the magnetic field on the space-time geometry, i.e. the non-linearity of the interaction between gravitation and magnetism. One manifestation of this would be the possibility of the magnetic field inducing torsion in space-time by inducing an effective spin-density in the collapsing matter. Eqs.(2.8) - (2.14) above, then would imply that a given magnetic field B distributed through the collapsing matter would give rise to an effective spin-density

$$\sigma_{eff} = (3Bc/8\pi)(1/2\alpha G)^{1/2} \qquad (2.30)$$

corresponding to an effective torsion

$$Q = 4\pi G\sigma_{eff}/c^3 = (3B/2c^2)(1/2\alpha)^{1/2}G^{1/2} =$$
$$= (3B/2c^2)(G//2\alpha)^{1/2} \qquad (2.31)$$

This corresponds to a length scale Q , thus implying that the effects of torsion induced by the magnetic field become significant in a collapsing body of instantaneous radius R_I , if

$$B > (2c^2/R_I)(2\alpha/G)^{1/2} = B_c \qquad (2.32)$$

For typical neutron star sizes, this corresponds to $B \approx 10^{16}$ Gauss, which shows that in a neutron star accreting matter to form a black hole, (one of the proposed scenarios for formation of binary black hole), the magnetic field could prevent collapse to a singular point.

We would have to consider a torsion modified Poisson equation of the form:

$$\nabla^2\Phi = 4\pi G(\rho - G\sigma^2) \qquad (2.33)$$

which has a solution having a minimum finite radius. We notice that the non-linear gravitational coupling in the second term of the brackets has an opposite sign to the gravitationally attractive first term, and moreover goes as the square of the density. The two terms become comparable for $B > B_c$. So the above argument on the ineffectiveness of the magnetic field, based on non-coupling of the magnetic field to torsion and geometry, no longer holds and it is possible to form an object of finite size supported by the torsion induced by the magnetic field on the collapsing geometry (see also Chapter VII).

4. Gravimagnetism and torsion

Many arguments are given in support of the existence of a gravimagnetic field in nature. We like also to see if torsion can give some new effects.

The formal analogy between Coulomb's law of electrostatic

force between two charges, i.e. $q_1q_2/4\pi\varepsilon_o r^2$ and the Newtonian law of gravitational force between two masses i.e. Gm_1m_2/r^2, suggests that the analogous quantity for electrical permittivity (ε_o) in gravitation is $(1/4\pi G) = \varepsilon_G$.

Since any field theory of gravity would require the existence of finite velocity of propagation of gravitational influences, i.e. gravitational waves moving at light velocity, one can deduce by analogy with electromagnetism (where c = velocity of electromagnetic waves = $1/\sqrt{\varepsilon\mu}$),that the corresponding gravitational permittivities and permeabilities are related by $c^2 = 1/\varepsilon_G\mu_G$. This implies that the gravitational permeability must be given by $\mu_G = 4\pi G/c^2$ [21], and this would be the coefficient of a gravimagnetic force, which is velocity dependent.

Many other arguments can be given to justify such a force with this value of μ_G. For instance, if one considers stationary masses and charges such that the force of electrostatic repulsion balances that of gravitational attraction and then,while maintaining their distance, imparts to them a uniform velocity v, this would tend to disturb the equilibrium by inducing a magnetic force, so that to maintain equilibrium one must invoke a magnetic type gravitational force dependent on the relative velocity between the bodies which is of the form [22],[23]: ~ $Gm_1m_2v^2/c^2r^2$, so that we have a coefficient ~ G/c^2 analogous to permeability μ for the magnetic field B.

The equations of motion in general relativity already contain gravimagnetic forces. For instance the geodesic equation to first order in h are [24]:

$$c^2 \frac{d^2x^i}{d\,s^2} = \left(\frac{d\,x^o}{d\,s}\right)^2 \left[\nabla \left(c^2 h_{oo}/2\right) - \partial(ch_{oi})/\partial t\right] + \qquad (2.34)$$

$$+ c^2 \frac{d\,x^o}{d\,s}\frac{d\,x^j}{d\,s}\left[h_{jo,i} - h_{io,j}\right] - c^2 h_{ij,o}\frac{dx^j}{ds}\frac{dx^o}{ds} - \Gamma^i_{jk} v^i v^k$$

If one uses this to study the motion of a particle inside a hollow spherical shell, the geodesic equation (with $h_{\mu\nu}$ related to the gravitational potentials as $h_{\mu\nu} = \Phi_{\mu\nu} - (1/2)\delta_{\mu}\Phi$) takes the form:

$$c^2 \frac{d^2\bar{x}}{d\,s^2} = \nabla V - \left(\frac{dx^o}{d\,s}\right)^2 \frac{\partial \bar{A}}{\partial t} + \bar{v} \wedge \nabla \wedge \bar{A} \qquad (2.35)$$

with $V = c^2 h_{oo}/2$, $A_i = ch_{oi}$. This is strictingly similar (for a time independent $\bar{A}$) to the Lorentz force $q\bar{E} + (\bar{v}/c)\wedge\bar{B}$ in electromagnetism, with $\bar{\nabla} \wedge \bar{A}$ analogous to the gravimagnetic field B_G (we have $\bar{E}_G = \nabla V$, $\bar{B}_G = \bar{v} \wedge \nabla \wedge \bar{A}$). Thus a relativistic theory of gravity inevitably leads to magnetic type of gravitational forces.

In the Einstein-Cartan theory, the torsion tensor Q_{ijk} (which is the skew-symmetrical part of the generalized affine connection $\Gamma^k_{[ij]}$) is related to the spin density tensor $J_{ij}^{\ k}$ (also acting as a source term through the relations (see eq.(2.10)):

$$Q_{ij}{}^{k} = (8\pi G/c^2)(J_{ij}{}^{k} - (1/2)\delta^k_i J_{bj}{}^{b} - (1/2)\delta^k_j J_{ib}{}^{b}) \qquad (2.36)$$

In general in a polarized medium, with N aligned spins per unit volume, the torsion is given by [25,26]:

$$\bar{Q} = (4\pi G/c^2)\bar{S}\ N \qquad (2.37)$$

(where Q is the spatial part of Q^α, see eq.(2.41)).

So we have the formal analogy of the interaction energy between spin and torsion, $\bar{S}\cdot\bar{Q}$, with the interaction energy of magnetic dipole moment in magnetic field, i.e. $\bar{\mu}\cdot\bar{H}$. The coefficient of 'permeability' is again $4\pi G/c^2$! from different considerations! In general for spinning matter we will have a gravitational permeability different from that of free space. We know that electrostatic forces can be drastically different in mareials with different ε and μ. There is a possibility of similar kind of shielding for gravitational forces, although very weak. This is inevitable once we have a dependence on ε and μ which are density dependent.

We know in electromagnetism that ε depends on density of medium. For instance in a plasma or metal, $\varepsilon = 1 - \omega_p^2/\omega^2$, i.e. frequency dependent. Here $\omega_p^2 = 4\pi ne^2/m_e$, n is the number density of charge carriers. The propagation of waves with $\omega \approx \omega_p$ is drastically affected, leading to a shielding or damping effect. For the gravitational case, the analogous quantity (in the case of propagating torsion) is

$$\omega_G^2 = Gn_o m_n, \qquad (2.38)$$

n_o = number density of particles of mass m_n. For ordinary matter $n_o \approx 10^{24}$/cc , $m_n = 10^{-24}$g. Then

$$\omega_G \approx 10^{-4} \text{ Hz} \qquad (2.39)$$

while for neutron stars

$$\omega_G \approx 10^{4} \text{ Hz} \qquad (2.40)$$

As a collapsing star emits gravitational waves with $\omega \approx 10^4$ Hz, this could drastically affect the propagation of such waves through the dense neutron star medium, like metals for electromagnetic waves. In particular all gravitational waves with frequency $\geq 10^4$ Hz would be reflected within the star and would not propagate outside, so one must take into account the consequences of this fact when considering high frequency gravitational radiation.

In the case of the Moon we have $n_O \sim 10^{23}/\text{cm}^3$ and $\omega_G \sim 10^{-4}$ Hz for interplanetary matter $\omega_G \sim 10^{-15}$ Hz and $n_o \sim 10/\text{cm}^3$. As all the gravitational waves emitted by sun or in the solar system (Earth-Sun $\omega \approx 10^{-7}$ Hz, Earth-Moon $\omega \approx 10^{-6}$ Hz) have frequency $<< 10^{-4}$ Hz, this would have negligible effect on the absorption by the Moon, unlike in the case of neutron stars where we have complete absorption.

One should also have an analog of Faraday rotation. The fact that ε and therefore μ depends on density, however weakly, suggests that different materials with different densities should affect the propagation of gravitational effects through them in different ways. The analog of the gyrofrequency $eB/m = \omega_B$ here is $\omega_B \sim \bar{S}\cdot\bar{Q}/\hbar$. So one can evaluate the Faraday rotation in this case. The torsion has a dipolar form [21]:

$$Q^{\alpha} \cong (9/4)(G/c^2)(1/r^3)\text{const} \qquad (2.41)$$

(where $Q^{\alpha} = (1/3!)\eta^{\alpha jik}Q_{jik}$
The gravitational permeability $\sim 4\pi G/c^2$ suggests a propagating torsion field of form $Q_{\mu} = \delta_{\mu}\Phi$ and if spins are partially polarized in a preferred direction, a dipolar torsion field is produced of form [27]:

$$\bar{Q} = \nabla\Phi = (G/c^3r^5)\left[\bar{S}\ r^2 - 3\ \bar{x}\ (\bar{S}\cdot\bar{x})\right] \qquad (2.42)$$

This is different from Lense-Thirring effect which involves the rotation of the frame of reference. Here we have interaction between spins similar to that in ferromagnetism [9]. In fact the similarity with the Langevin formula for susceptibility ($\kappa = M/H$) of a ferromagnetic has been obtained. So the difference from the usual gravimagnetic field of General Relativity is clear.

This is just of the form of a magnetic dipole field with a $\sim 1/r^3$ dependence! and we have also the right coefficient G/c^2 expected from gravitational permeability. So this is just a gravitational magnetic dipole field and it could give substantial contribution to neutron star magnetic fields which are also dipolar with a ~ $1/r^3$ dependence.

We can also consider the gravitational action to be made up of two parts: an 'electric' part and a 'magnetic' part. The usual Hilbert action would be the 'electric' part, i.e. with 'dielectric constant 1/G i.e. $(1/G)\int R\, d^4x$. For the 'magnetic' part we can have the torsion Lagrangian

$$\sim \quad (4\pi G/c^2)\int \bar{\psi}\, \gamma_\mu (1+\gamma_5)\psi\ \psi\, \gamma_\mu (1+\gamma_5)\bar{\psi} \quad \text{etc.} \tag{2.43}$$

(for torsion between two fermions) so we have the permittivity 1/G and the permeability $4\pi G/c^2$ appearing in the two parts respectively. The torsion part would play the role of 'magnetism' as far as gravitational forces are considered. The analogue of Faraday rotation gives

$$\nabla\vartheta = \frac{2\pi mc}{\hbar}\frac{G\lambda^2}{c^4}\int_0^D n\ (\bar{S}\cdot\bar{Q})\ dS \qquad \text{radians} \tag{2.44}$$

This formula is obtained by analogy with electromagnetism. In fact we have in the case of e.m. that the propagation vector is [28]

$$K_{RL} \cong \frac{\omega}{c}\left[1 - \frac{\omega_p^2}{2\bar{\omega}^2}\left(1 \pm \frac{\omega_B}{\omega}\right)\right] \tag{2.45}$$

so in our case when torsion is introduced (being $\omega^2_{p_{grav}} = 4\pi G n m_0$ and $\omega_{B_{grav}} = (\bar{S}\cdot\bar{Q})/\hbar$) :

$$K_{RL} \cong (\omega/c)\left[1 - \frac{4\pi G n m_0}{2\omega^2}\left(1 \pm \frac{\bar{S}\cdot\bar{Q}}{\hbar\omega}\right)\right] =$$

$$= (\omega/c)\left[1 - \frac{2\pi G n m_0}{\omega^2} \pm \frac{2\pi G n m_0 \bar{S}\cdot\bar{Q}}{\hbar\omega^3}\right] \qquad (2.46)$$

so that analogously to

$$\frac{d\vartheta_P}{d\omega} = -\frac{4\pi e}{c n \omega^3}\int_0^D n\, ds \qquad (2.47)$$

and then anolgous to the Faraday rotation

$$\Delta\vartheta_B = \frac{2\pi e^3\lambda^2}{m^2c^4}\int_0^D n B\, ds \qquad (2.48)$$

we have in the gravitational case (with torsion):

$$\Delta\vartheta = \frac{2\pi m c}{\hbar}\frac{G\lambda}{c^4}\int_0^D n\,(\bar{S}\cdot\bar{Q})\, ds \qquad (2.49)$$

References

[1] P.M.S.Blackett - Nature 159, 658 (1947)
[2] D.A.Ahluwalia and T.Y.Wu - Lett.Nuovo Cimento 23, 406 (1978)
[3] S.P.Sirag - Nature 278, 358 (1979)
[4] P.Brosche - in "Spin, Torsion, Rotation and Supergravity" ed. by Peter G.Bergmann and Venzo de Sabbata, Plenum Press, NATO, ASI Series B Physics Vol.58, N.Y.1980 pg.375
[5] P.S.Wesson - Physics Rev. D23, 1730 (1981)
[6] V.de Sabbata and M.Gasperini - Lettere Nuovo Cimento 38, 93 (1983)

[7] V. de Sabbata and M.Gasperini - Lettere Nuovo Cimento 27, 133 (1980)

[8] V.de Sabbata nad M.Gasperini - in "Origin of Galaxies" ed.by Venzo de Sabbata, World Sci.Singapore 1981 pg.181

[9] V.de Sabbata and P.Rizzati - Lettere Nuovo Cimento 20, 525 (1977)

[10] V.de Sabbata - in "Gravitational Measurements, Fundamental Metrology and Constants" ed. by Venzo de Sabbata and V.N.Melnikov, Kluwer Acad.Publ. 1988 pag.115

[11] V.de Sabbata and C.Sivaram - Astrophys.and Space Sci. 158, 347 (1989)

[12] P.A.M.Dirac - Proc.Roy.Soc.A338, 439 (1974)

[13] P.A.M.Dirac - Proc.Roy.Soc.A165, 199 (1938)

[14] P.M.S.Blackett - Philos.Trans.Roy.Soc.London Ser. A245,309 (1952)

[15] V.de Sabbata and M.Gasperini - Lett.Nuovo Cimento 25, 489 (1979)

[16] V.de Sabbata and C. Sivaram - Il Nuovo Cimento 102, 107 (1988)

[17] A.H.Guth - Phys.Rev.D,23, 347 (1981)

[18] C.Isham, A.Salam and J.Strathdee - Nature Phys.Sci. 244, 82(1973)

[19] V.de Sabbata and M.Gasperini - Lett.Nuovo Cimento 27, 289 (1980)

[20] E.Asseo and H.Sol - Phys.Reports 148, 307 (1987)

[21] V.de Sabbata and M.Gasperini - "Introduction to Gravitation" World Scientific Singapore, 1985

[22] Wienfield W.Salisbury and Donald H.Menzel - Nature 252,664 (1974)

[23] Robert L.Forward - Amer.Jour.Phys. 31, 166 (1963)

[24] J.Weber - "General Relativity and Gravitational vawes" Interscience Publ. New York, 1961 pg 157

[25] V.de Sabbata and M.Gasperini - Phys.Rev. D23, 2116 (1981)

[26] V.de Sabbata and M.Gasperini - Lett.Nuovo Cimento 30, 503 (1981)

[27] V.de Sabbata and C.Sivaram - Il Nuovo Cimento 101, 273 (1989)

[28] J.D.Jackson - Classical electrodynamics, John Wiley, New York 1962 pg.227

CHAPTER III

TORSION AND MAXWELL EQUATIONS

1. General considerations

It has been repeatedly stated [1], in the framework of the Einstein-Cartan theory, that the electromagnetic field cannot be coupled to torsion, in order to preserve local gauge invariance.

Therefore the 'minimal coupling' principle i.e. the replacement of the flat metric $\eta_{\mu\nu}$ and of the partial derivative δ_μ with the covariant metric and derivative $g_{\mu\nu}$ and ∇_μ , cannot be applied, in the presence of torsion, to the special relativistic photonic Lagrangian; more generally, any Abelian or non Abelian gauge field cannot be minimally coupled to the gèometry of a Riemann-Cartan space.

The result finds a natural justification in the framework of the Poincaré gauge field theory of gravitation.

Several attempts have been made in order to apply minimal coupling without breaking gauge invariance: they are successfull, however, only at the cost of imposing arbitrary geometrical constraints upon torsion [2], or introducing a modified definition of gauge transformation, and allowing the propagation of torsion in vacuum [3].

Therefore these attempts lie beyond the limits of the Einstein-Cartan theory, where torsion is determined only by spin densities and cannot propagate, but it is rigidly bound to the matter distribution in space-time.

In order to preserve the gauge invariance in the Einstein-Cartan theory, the most simple hypothesis is to assume that

"photons neither produce nor feel torsion" (3.1)

i.e.the Maxwell equations in a Riemann-Cartan space U_4 are identical to the same equations written in the Riemann space V_4 obtained from U_4 putting torsion equal to zero.

We like to emphasize that the statement (3.1) can be assumed to be strictly valid only as long as the electromagnetic (e.m.) field is treated as a classical, not quantized field. When a quantum point of view is adopted, however, (3.1) is recovered only as a first approximation, since, in general, according to a quantum description of the e.m. field in a space with torsion, one may always expect an interaction between the photons and the torsionic background.

In fact a photon, with a process of the second order in the perturbative development of the e.m. interaction, can virtually disintegrate into an electron-positron pair (vacuum polarization effect); since these particles are massive fermions, which couple to torsion, they feel the presence of a torsionic background; as a consequence, also the e.m. field is affected by torsion (see Fig.3.1); even if torsion does not directly interact with the photon field, it does interact with the virtual pairs produced in vacuum by a "physical" photon. This interaction preserves the gauge invariance and the Maxwell equations are modified by a quantum contribution of the second order, so that, to the zeroth order, i.e. in the classical field approximation, the coupling with torsion disappears and we recover the usual form as the Maxwell equations.

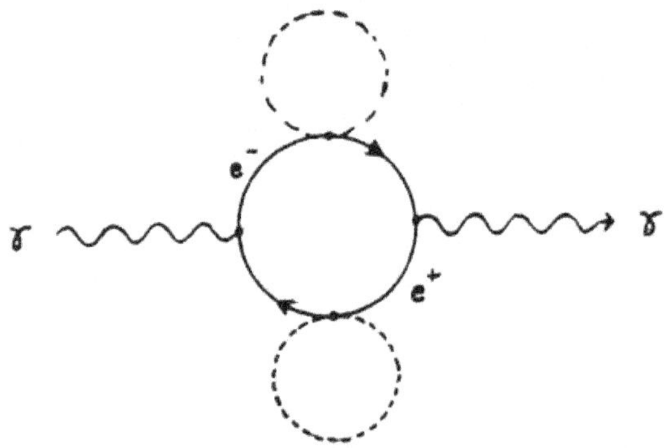

Fig.(3.1) The dashed lines represent the interaction of the virtual pair with the external torsionic background

We shall explicity compute the torsionic contributions, and then we shall formulate an empirical procedure, which provides the same results as the perturbative computation so that it can replace the minimal coupling in the case of the e.m. field. The modified Maxwell equations will be applied to evaluate the influence of the spin of a particle upon its own e.m. field. We shall compare the torsion contributions with the effects induced by the spin through a modification of the metric.

As a second application we shall suppose that torsion is produced by a cosmological neutrino sea and we shall consider the effects of the neutrino distribution upon the propagation of an e.m. wave. The space filled with neutrinos behaves like an optically active and dispersive medium, since it cause a rotation of the linear polarization and can be described by an index of refraction depending on the frequency.

The joint action of a gravitational field and of a torsionic sea can act as a prism, producing the spatial splitting of the waves with different wave-length contained in a wave packet.

2. Maxwell's Equations in a Riemann-Cartan space: a perturbative approach

In order to evaluate the torsion contribution to the Maxwell equations, we write the field equations for the e.m. potential A_μ in momentum space

$$(p^2\delta^\nu_\mu - p^\nu p_\mu)\ A^\mu(p) = 4\pi J^\nu(p) \qquad (3.2)$$

Considering the phenomena of virtual pairs production, they must be modified as follows:

$$(p^2\delta^\nu_\mu - p^\nu p_\mu) A^\mu = 4\pi(J^\nu + \Pi^{\nu\mu}A_\mu) \qquad (3.3)$$

where $\Pi_{\nu\mu}$ is the polarization tensor. Since in general it satisfies

$$\Pi^{\nu\mu}(p)\ p_\mu = 0 \qquad (3.4)$$

the equations(3.3) are invariant under the gauge transformation of the second kind

$$A_\mu(p) \Longrightarrow A_\mu(p) + p_\mu\Lambda(p) \tag{3.5}$$

and so the e.m. gauge invariance is preserved. Notice that the creation of virtual pairs induces an additional current density $J^\nu = \Pi^{\nu\mu}A_\mu$, and that the polarization tensor is essentially the proportionality constant that relates the induced current to the field intensity. In vacuum $\Pi_{\nu\mu}$ may be written in the form

$$\Pi_{\nu\mu}(p) = \Pi^{(0)}_{\nu\mu} = \Pi(p^2)(p^2\eta_{\nu\mu} - p_\nu p_\mu) \tag{3.6}$$

Therefore the vacuum polarization has no (linear) effect on an e.m. wave travelling in empty space, since a free radiation field satisfies $\Box A^\mu = 0$ and then we have $\Pi^{(0)}_{\nu\mu}A^\mu = 0$.

For a field coupled to a current, in the linear approximation we have from (3.2) and (3.6) $\Pi^{(0)}_{\nu\mu}A^\mu = 4\pi\Pi(p^2)J_\nu$ and then the vacuum polarization tensor gives rise to a modification of the e.m. coupling constant $e^2 \Longrightarrow e^2[1 + \Pi(p^2)]$.

If however the production of virtual pairs takes place in the presence of an external field, one has $\Pi_{\nu\mu} \neq \Pi^{(0)}_{\nu\mu}$, and the interaction of this field with the pairs provides a contribution to the Maxwell equations in addition to the charge renormalization effect.

Without enter in the details of the procedure, we notice only that the computation of the polarization tensor from the S-matrix expansion to order e^2 gives (see Fig.3.2)

$$\Pi_{\nu\mu}(p) = 0\ [ie^2/(2\pi)^4]\int d^4q\ \mathrm{Tr}\ [S_F(q)\gamma_\nu S_F(q - p)\gamma_\mu] \tag{3.6}$$

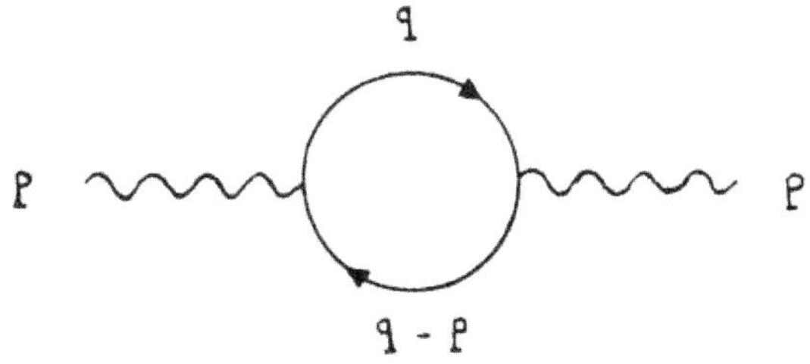

Fig.(3.2) Photon self-energy diagram

where $S_F(p)$ (see [4], [5], [6]) is the electron propagator in momentum space and in vacuum is

$$S_F^{(0)} = - i/(i\gamma_\mu p^\mu + m) \qquad (3.7)$$

and the vacuum polarization tensor $\Pi^{(0)}_{\nu\mu}(p)$ produces only a charge renormalization effect.

Following this procedure, writing the Dirac equation in U_4, the propagator in the presence of torsion and developing the modified propagator to the first order in the spin-torsion coupling constant (that we take equal to Newtonian constant G) i.e. $S_F = S_F^{(0)} + S_F^{(1)}$ where

$$S_F^{(1)} = [1/(i\bar{\gamma}\cdot\bar{p} + m)]i\bar{\gamma}\cdot\bar{Q}\gamma_5[1/(i\bar{\gamma}\cdot\bar{p} + m)] = S_F^{(0)}(-i\bar{\gamma}\cdot\bar{Q}\gamma_5)S_F^{(0)} \qquad (3.8)$$

we find

$$\Pi^{(1)}_{\nu\mu}(p) = i\ e^2/(3\pi^2)\eta_{\nu\mu\rho\sigma}Q^\rho p^\sigma \qquad (3.9)$$

being

$$Q^\rho = (1/16)\eta^{\nu\sigma\mu\rho}K_{\nu\sigma\mu} \qquad (3.10)$$

(where $K_{\nu\sigma\mu}$ is the usual contorsion tensor).

Notice that, as previously stated, $\Pi^{(1)}_{\nu\mu} p^{\mu} = 0$ and the gauge invariance is preserved.

Taking into account the torsion contribution (3.9), the equation (3.3) becomes, in coordinate space ($ip_{\mu} \Longrightarrow \partial_{\mu}$) and putting $F_{\nu\mu} = 2\, \partial_{[\nu}A_{\mu]}$:

$$\left.\begin{aligned} \partial_{\mu}F^{\nu\mu} &= 4\pi J^{\nu} + (2\alpha/3\pi)\ \eta^{\nu\rho\sigma\mu} Q_{\sigma} F_{\rho\mu} \\ &\text{which, together with} \\ \partial_{[\nu}F_{\sigma\mu]} &= 0 \end{aligned}\right\} \quad (3.11)$$

constitute the Maxwell equations in the presence of torsion (α is the fine-structure constant $\alpha = e^2/\hbar c$).

Now this photon-torsion interaction obtained through an explicit perturbative calculation considering the virtual pairs production, that is the vacuum polarization effect, can be translate in the following, so to say, macroscopical empirical procedure replacing the minimal coupling principle and that we call 'semi-minimal' because is minimal at microscopical level:

1) Apply the minimal coupling principle only to the field tensor, and not to the potential vector; then $F_{\nu\mu} = 2\partial_{[\nu}A_{\mu]}$ and $\partial_{[\nu}F_{\sigma\mu]} = 0$ still hold, but the other equation becomes

$$\nabla_{\mu}F^{\nu\mu} \equiv F^{\nu\mu}{}_{|\mu} + K_{\mu\rho}{}^{\nu}F^{\mu\rho} + K_{\mu\rho}{}^{\mu}F^{\rho\nu} = 4\pi J^{\nu} \qquad (3.12)$$

where a bar denotes covariant derivative involving the Christoffel symbols

2) Use for the coupling only the totally antisymmetric part of the torsion, that is

$$K_{\nu\sigma\mu} = -(2/3)\eta_{\nu\sigma\mu\rho}Q^{\rho} \qquad (3.13)$$

3) Replace the spin-torsion coupling constant G by $G' = \alpha G/\pi$.

By applying these rules, we obtain the following electromagnetic equations:

$$\left.\begin{aligned} &\partial_{[\nu}F_{\sigma\mu]} = 0 \\ &[1/(-g)^{1/2}]\partial_{\mu}[(-g)^{1/2}F^{\nu\mu}] = 4\pi J^{\nu} + (2\alpha/3\pi)\eta^{\nu\mu\rho\sigma}F_{\mu\rho}Q_{\sigma} \end{aligned}\right\} \qquad (3.14)$$

which in the case of a flat metric are reduced exactly to (3.11).

In this case the rules 1), 2), 3) constitute the new procedure which we call "semi-minimal" because it is minimal at a microscopical level, i.e. when considering the interaction of the $e^- e^+$ pairs with torsion, even if it is non minimal macroscopic photon-torsion interaction.

The things that so far we have said are valid in the Einstein-Cartan theory, but we can also introduce a propagating torsion, imposing that the axial-vector part of the torsion tensor (that is (3.10)) is determined by the pseudoscalar potential φ according to $Q_{\beta} = \partial_{\beta}\varphi$ so that we have the Lagrangian

$$\mathcal{L}_{I} = \alpha\ G\ \varepsilon^{\mu\nu\alpha\beta}A_{\mu}F_{\nu\alpha}\partial_{\beta}\varphi \qquad (3.15)$$

where α is the fine structure constant, G the gravitational constant, $\varepsilon^{\mu\nu\alpha\beta}$ the totally antisymmetric tensor, $F_{\mu\nu} = \partial_{\mu}A_{\nu}$ -

$\partial_\nu A_\mu$ and $\partial_\mu\varphi$ is the axial-vector part of the torsion tensor determined by the pseudoscalar potential φ [7].

As a consequence, we can observe that an electric charge interacting with a constant torsionic background, can produce a static magnetic field even in its rest system [8].

Consider in fact a pointlike charge distribution $\rho(x) = e\partial(x)$ placed in an external torsionic background generated by a constant semi-classical spin distribution [9] at rest with the charge density. We have then $Q_\mu u^\mu = 0$ where u^μ is the four velocity vector of the charged particle. Then the modified Maxwell equations (3.11) in the charge rest system where $\partial H/\partial t = \partial E/\partial t = 0$ $J^\mu \equiv (0, c\rho)$ and $Q^\mu \equiv (\bar{Q}, 0)$ are reduced ($E_k = F_{k0}$, $H_k = (1/2)\varepsilon_{kll}F_{ll}$ where the latin indices run from 1 to 3):

$$\bar{\nabla}\cdot\bar{E} = 4\pi\rho + (4\alpha/3\pi)(\bar{Q}\cdot\bar{H}) \, , \qquad \bar{\nabla}\cdot\bar{H} = 0$$
$$\bar{\nabla}\wedge\bar{E} = 0 \, , \qquad \bar{\nabla}\wedge\bar{H} = (4\alpha/3\pi)\bar{E}\wedge\bar{Q} \tag{3.16}$$

that is

$$\nabla^2 E = 4\pi\nabla\rho + (4\alpha/3\pi)\nabla(\bar{H}\cdot\bar{Q})$$
$$\nabla^2 H = (4\alpha/3\pi)Q\left(4\pi\rho + (4\alpha/3\pi)H\cdot Q\right) \tag{3.17}$$

From these equations we get to the first order in the torsion contributions

$$\bar{E}(\bar{x}) = -\bar{\nabla}\varphi \, , \qquad \bar{H}(\bar{x}) = -(4\alpha/3\pi)\varphi\,\bar{Q} \tag{3.18}$$

where $\varphi = e/r$ is the electromagnetic potential.

The constant torsionic background induces then a static magnetic field, spherically symmetrical [$H = H(r)$] and aligned along the torsion direction. If torsion is produced by the spin of the particle itself then, since a charged and spinning particle can be regarded as a sort of 'cloud ' with a spatial extension ~ $r_0 = \hbar/mc$ and with an 'electromagnetic radius' $r_{e.m.} = e^2/mc^2$, which is always smaller than r_0 ($r_{e.m.} = \alpha r_0 \sim 10^{-2} r_0$), it is justified to assume a pointlike charge distribution and a nearly constant spin density Σ and then use the results (3.18) in order to evaluate the torsion contribution to the e.m. field inside the Compton radius r_0.

In the case of a Dirac particle putting $\Sigma = \hbar/2V$ where $V = 4\pi r_0^3/3$, and supposing that the spin is aligned along the x_3 axis, we have

$$Q^\alpha \simeq (9/4)(G\hbar/c^3 r_0^3)\delta_3^\alpha \qquad (3.19)$$

The magnetic field induced by torsion becomes then ($H_1 = H_2 = 0$)

$$H_3 = -\ (3\alpha/\pi)(G\hbar/c^3 r_0^3)(e/r) \qquad (3.20)$$

and it must be added to the usual dipolar field produced by the particle magnetic moment $\mu = e\hbar/2mc$.

In spite of the fact that the torsion contribution (3.20) must vanish outside the Compton radius (since $Q^\alpha = 0$ for $r > r_0$), it is worth noticing that it influences the total magnetic field also outside this radius, because of the matching conditions between internal and external field across the spherical Compton surface of discontinuity.

In the same way as the gravitational field outside spinning matter, in the Einstein-Cartan theory is produced by an "effective" source containing also the torsion contribution, besides the usual gravitational terms, also the total "external" magnetic field of the spinning particle is produced by an effective magnetic dipole $\mu' = \mu + \Delta\mu$ where, from (3.20) is:

$$\Delta\mu \simeq -(\alpha/\pi)(Gme/c^2) \quad (3.21)$$

Of course this correction is fully negligible, unless one assumes, in the case of hadrons, that Newtonian gravity is dominated by strong gravity: then the Einstein-Cartan field equations must be rewritten in terms of the strong metric $f_{\mu\nu}$ and the hadronic spin is the source of a 'strong torsion tensor' [10], [11].

In this case one must replace in (3.21) the Newton constant G by the strong-coupling constant $G_f \simeq \hbar c/m_p^2$ (m_p is some typical hadron mass). The torsion contribution to anomalous magnetic moment is then $|\Delta\mu| \simeq \alpha\mu/\pi$ and it turns out to be of the same order as the correction due to the e.m. interaction.

But as regards the magnetic moment, we can also have, with the consideration of torsion, a magnetic moment for massless and/or uncharged particles.

In fact considering a spin-torsion coupling for neutrinos in the early universe we can show that a neutrino magnetic moment can be present also for zero-mass neutrinos [12]. We find the result

$$\mu_\nu = 10^{-21}\mu_B \quad (3.22)$$

which is of the same magnitude as that obtained in the

Salam-Weinberg electroweak theory.

In order to arrive at the result (3.22), first of all we have shown that torsion can induce flavour oscillations even for zero mass neutrinos [13], the torsion interaction inducing an energy splitting giving rise to an oscillation length [14], [15]:

$$l_T = 2\pi/\Delta E_T = c^3/2G_T\hbar\rho\delta^2 \qquad (3.23)$$

where, owing to the fact that the torsionic contact interaction Lagrangian between two spin-half particles of wanishing mass is formally identical [16] to the weak-interaction Lagrangian, and may be written in (V - A) form, the spin-torsion coupling constant G_T has the same magnitude of the Fermi interaction constant; ΔE_T is $= E_{1T} - E_{2T} = 2(k_1 - k_2)\bar{Q}\cdot\bar{S}$ with k_1 and k_2 dimensionless real parameters; $\delta^2 = |k_1 - k_2|$. Then (3.23) gives $l_T \approx 10^{-4}$cm implying an oscillation time scale significantly shorter then the diffusion time-scale.

The point is that torsion interaction can induce magnetic moment [8] in addition to a magnetic field as in ref.[17], and, moreover, this effective magnetic moment would be induced by torsion even for zero mass neutrinos just as with the case of oscillations [13]. In this manner we arrive at the result (3.22) where (see [12]) μ_ν is temperature dependent. The result (3.22) was obtained for the corresponding temperature T in the early universe ($T \approx 10^{10}$K), but if we take into account (see Chapter VI) that we have a time-dependent G_T falling as

t^{-1}, this would compensate for T^2 going as t^{-1}, so that μ_ν would remain more or less constant with the value $\mu_\nu \approx 10^{-21}\mu_B$ acquired in the early universe.

References

[1] B.Kuchowicz - Acta Phys.Cosm. 3, 109 (1975)

[2] M.Novello - Phys.Lett.59A, 105 (1976)

[3] S.Hojman, M.Rosenbaum and M.P.Ryan - Phys.Rev.D17, 3141 (1978)

[4] V.de Sabbata and M.Gasperini - Lett.Nuovo Cimento 28, 229 (1980)

[5] V.de Sabbata and M.Gasperini - Phys.Rev.23D, 2116 (1981)

[6] V.de Sabbata and M.Gasperini - Physics Letters 83A, 115 (1981)

[7] V.de Sabbata and M.Gasperini - Lett.Nuovo Cimento 30, 193 (1981)

[8] V.de Sabbata and M.Gasperini - Lett.Nuovo Cimento 28, 234 (1980)

[9] W.Adamowicz and A.Trautman - Bull.Acad.Pol.Sci. 23, 339 (1975)

[10] V.de Sabbata and M.Gasperini - Gen.Rel.Grav.10, 825 (1979)

[11] V.de Sabbata and M.Gasperini - Lett.Nuovo Cimento 27, 289 (1980)

[12] V.de Sabbata and C.Sivaram - Il Nuovo Cimento 105B, 603 (1990)

[13] V.de Sabbata and M.Gasperini - Il Nuovo Cimento 66A, 479 (1981)

[14] V.de Sabbata and C.Sivaram - Il Nuovo Cimento 101A, 273 (1989)

[15] V.de Sabbata and C.Sivaram - Ann.der Phis.7, 627 (1989)

[16] V.de Sabbata and M.Gasperini - Gen.Rel.Grav.10, 731 (1979)

[17] V.de Sabbata and C.Sivaram - Il Nuovo Cimento B102, 107 (1988)

CHAPTER IV

MAXWELL ELECTRODYNAMICS IN SPACE WITH TORSION

1. The effect of torsion on electromagnetic field of a spinning charge at rest

Now starting from eqs.(3.14) we want to evaluate, in a first approximation, the effects produced by the spin of a particle upon its own e.m. field. First of all rewrite the Maxwell eqs.(3.14) in a 3-dimensional form. Putting:

$$\left.\begin{array}{ll} E_k = F_{k0}\,, & B_k = (1/2)\varepsilon_{kil}F_{il} \\ D_k = (-g)^{1/2}F^{0k}\,, & H_k = (1/2)\varepsilon_{kil}F^{il} \end{array}\right\} \quad (4.1)$$

where $F^{il} = g^{im}g^{ln}F_{mn}$ and ε_{kil} is the 3-dimensional Levi-Civita symbol (latin indices run from 1 to 3) we get

$$\left.\begin{array}{l} \bar{\nabla}\cdot\bar{D} = 4\pi\rho + (4\alpha/3\pi)(-g)^{1/2}\bar{Q}\cdot\bar{B}\ ; \quad \bar{\nabla}\cdot\bar{B} \\ \bar{\nabla}\wedge\bar{E} = -\ (1/c)(\partial\bar{B}/\partial t)\ ;\ \bar{\nabla}\wedge\bar{H} = (4\pi/c)\bar{J} + (1/c)(\partial\bar{D}/\partial t)\ + \\ \qquad + (4\alpha/3\pi)(-g)^{1/2}(Q^0\bar{B} + \bar{E}\wedge\bar{Q}) \end{array}\right\} \quad (4.2)$$

Now if we want evaluate the effects produced by the spin of a particle upon its own e.m. field we must solve eqs.(4.2) taking as source of torsion the spin of the particle itself.

In the case of a Dirac particle we have [1] (see also the previous chapter where a analogous calculation was made in the case of flat space):

$$Q^\alpha = -\ 3\pi(G\hbar/c^3)\bar{\psi}\ \gamma_5\gamma^\alpha\psi \quad (4.3)$$

and then in the rest system of the particle where $J^{\alpha} \equiv (\bar{0}, c\rho)$, $\partial\bar{B}/\partial t = 0 = \partial\bar{D}/\partial t$, we can put (in our approximation)

$$Q^0 = 0 \qquad Q^k = (9/4)(G\hbar/c^3)(1/r_0^3)\delta_3^k \tag{4.4}$$

(we have supposed the spin aligned along the x_3 axis)

The equations (4.2), putting $g_{\alpha\beta} = \eta_{\alpha\beta}$ since we want to evaluate the specific torsion contributions, become ($D = E$, $B = H$)

$$\left.\begin{array}{ll} \bar{\nabla}\cdot\bar{E} = 4\pi\rho + (4\alpha/3\pi)\bar{H}\cdot\bar{Q}\ ; & \bar{\nabla}\cdot\bar{H} = 0 \\ \bar{\nabla}\wedge\bar{E} = 0\ ; & \bar{\nabla}\cdot\bar{H} = (4\alpha/3\pi)\bar{E}\wedge\bar{Q} \end{array}\right\} \tag{4.5}$$

that is

$$\left.\begin{array}{l} \nabla^2\bar{E} = 4\pi\bar{\nabla}\rho + (4\alpha/3\pi)\bar{\nabla}(\bar{H}\cdot\bar{Q}) \\ \nabla^2\bar{H} = (4\alpha/3\pi)\bar{Q}\left(4\pi\rho + (4\alpha/3\pi)\bar{H}\cdot\bar{Q}\right) \end{array}\right\} \tag{4.6}$$

(see also (3.17)) from which we get, neglecting second order torsion terms

$$\bar{E} = e\bar{x}/r^3\ ; \qquad \bar{H} = -(4\alpha/3\pi)(e/r)\bar{Q} \tag{4.7}$$

where $r = (x_k x^k)^{1/2}$, and $\bar{Q}$ is given by (4.4).

Therefore the effect of torsion is to induce an axially symmetrical magnetic field, aligned along the spin direction. As said in the previous chapter this term must be added to the usual dipolar magnetic field, so that it may contribute to the well known anomaly in the magnetic moment of the elementary particles.

It is interesting to compare this 'internal' field with the 'external magnetic field due to the spin influence on the metric, outside the radius r_0 (Compton radius) where there is no torsion contribution, since $Q^{\alpha} = 0$.

In the weak field approximation, assuming the metric

$$g_{00} = -1 + 2\varphi \quad ; \quad g_{ik} = \delta_{ik}(1 + 2\varphi)$$
$$g_{0i} = (2G/r^3c^3)\varepsilon_{ikl}x^k\sigma^l \quad ; \quad (-g)^{1/2} \simeq 1 + 2\varphi \qquad (4.8)$$

where $\varphi = Gm/rc^2$ is the gravitational potential and $\bar{\sigma}$ is the spin of the particle, we get from (4.1)

$$\left.\begin{aligned} \bar{D} &= \bar{E}\,(1 + 2\varphi) + (2G/r^3c^3)\bar{H}\wedge(\bar{x}\wedge\bar{\sigma}) \\ \bar{B} &= \bar{H}\,(1 + 2\varphi) + (2G/r^3c^3)(\bar{x}\wedge\bar{\sigma})\wedge\bar{E} \end{aligned}\right\} \qquad (4.9)$$

The eqs.(4.2), in the absence of torsion and in the rest system of the particle become

$$\bar{\nabla}\cdot\bar{E}\,(1 + 2\varphi) = 4\pi\rho + (2G/c^3)\bar{\nabla}\cdot\left[(\bar{\sigma}/r^3)(\bar{x}\cdot\bar{H}) - (\bar{x}/r^3)(\bar{\sigma}\cdot\bar{H})\right]$$
$$\bar{\nabla}\cdot\bar{H}\,(1 + 2\varphi) = (2G/r^3)\bar{\nabla}\cdot\left[(\bar{x}/r^3)(\bar{E}\cdot\bar{\sigma}) - (\bar{\sigma}/r^3)(\bar{E}\cdot\bar{x})\right] \qquad (4.10)$$

and they give, neglecting terms of higher order than first in G

$$\bar{E} = (e\bar{x}/r^3)\left(1 - (2Gm/r^2)\right)$$
$$\bar{H} = (2G/c^3)(e/r^6)[\bar{x}(\bar{\sigma}\cdot\bar{x}) - \bar{\sigma}r^2] \qquad (4.11)$$

Notice that also through its metric influence, the spin induces a magnetic field in the particle's rest system, and that this field is stronger by a factor α^{-1} than the one induced directly by torsion, which vanishes outside the Compton radius, (this could not be true any longer, by the way, if the spin-torsion coupling constant is different from the mass-curvature constant G). The coupling between torsion and the e.m. field plays however a fundamental role in those space with a metric tensor whose deviations from $\eta_{\mu\nu}$ are negligible, and with a nonsymmetric affine connection.

This situation is not merely speculative since it could correspond, in nature, to the case of a space filled by a cosmological neutrino sea: it is well known, in fact, that neutrinos can produce torsion without affecting the metric of the space in which they live [2], [3].

It makes sense, then, to consider the Maxwell equations in a space with a given metric, to add, independently of the metric, the torsionic contributions, according to(4.2), in order to investigate the e.m. effects in the early universe (for instance induced by a cosmological neutrino sea).

2. Propagation of e.m. waves through a cosmological neutrino sea

a) rotation of linear polarization

The effects on electromagnetism of a cosmological neutrino sea have been investigated from the point of view interaction [4] and also of the photon-neutrini interaction postulated by Bandyopadhyay [5].

Since leptons, besides the weak, are also subject to the gravitational interaction, an hypothetical neutrino sea could influence the propagation of an e.m. wave also with the production of a torsionic background throughout the space.

We suppose that the neutrino sea has no preferred direction of spin orientation: this means that a frame exists, called the "rest frame of the sea" [4], [6] where the neutrino velocities are isotropically distributed and $Q^k = 0$. The eqs.(4.2) for an e.m. wave become then

$$\bar{\nabla}\cdot\bar{D} = 0 \quad ; \quad \bar{\nabla}\cdot\bar{B} = 0 \tag{4.12}$$

$$\bar{\nabla}\wedge\bar{E} = -(1/c)\partial\bar{B}/\partial t \quad ; \quad \bar{\nabla}\wedge\bar{H} = (1/c)\partial\bar{D}/dt + (4\alpha/3\pi)(-g)^{1/2}Q^0\bar{B}$$

where

$$Q^0 = -3\pi(G\hbar/c^3)\rho_\nu \tag{4.13}$$

and ρ_ν is the neutrino density i.e.the number of neutrinos minus the number of antineutrinos in a unitary volume. We suppose that ρ_ν is constant and uniform.

In order to investigate how is affected the e.m.radiation in this space, we consider a wave emitted by a spherical body of mass M, static, nonrotating and, for simplicity at rest with the neutrino background [7].

In the weak field approximation, we can use then the metric (4.8) with $\bar{\sigma} = 0$. From (4.1) we get

$$B = \varepsilon H \quad ; \quad D = \varepsilon E \quad ; \quad (-g)^{1/2} = \varepsilon \tag{4.14}$$

where $\varepsilon = 1 + 2GM/rc^2$ ($r = x^k x_k$ is the distance from the central body) and eqs.(4.12) are reduced to

$$\bar{\nabla}\cdot\bar{E} = -(\bar{E}\cdot\bar{\nabla}_\varepsilon/\varepsilon) \qquad \bar{\nabla}\cdot\bar{H} = -(\bar{H}\cdot\bar{\nabla}_\varepsilon/\varepsilon) \tag{4.15}$$

$$\bar{\nabla}\wedge\bar{E} = -(\varepsilon/c)(\partial\bar{H}/\partial t) \qquad \bar{\nabla}\wedge\bar{H} = (\varepsilon/c)(\partial\bar{E}/\partial t) - \eta\varepsilon^2\bar{H} \tag{4.16}$$

where $\eta = 4\alpha G\hbar\rho_n/c^3$.

Suppose that the e.m. radiation is a wave packet that leaves the body traveling along a radial direction which we choose to be the x^3-axis parallel to the neutrino current. If we confine our interest to the central part of the wave packet, for $x_1 \ll x_3 \gg x_2$, we have $\partial_1\varepsilon \ll \partial_3\varepsilon \gg \partial_2\varepsilon$ and we can approximate the radiation with a plane wave of components E_1, E_2, H_1, H_2 which are function of x_3 only. So we have satisfied (approximately) the divergence eqs.(4.15). The curl eqs. (4.16) yield two linearly independent solutions, representing a right- (E_+,H_+) and left-handed (E_-,H_-) wave; putting

$$E_\pm = E_1 + iE_2 = A_\pm e^{i(kx_3 - \omega_\pm t)} \tag{4.17}$$

from the modified Maxwell eqs.(4.16) we get the following dispersion relations for the two polarization modes

$$\omega_\pm^2 = (c^2/\varepsilon^2)k(k \mp \eta\varepsilon^2) \tag{4.18}$$

Notice that, in the absence of torsion ($\eta \Longrightarrow 0$), we have $\omega_+ = \omega_- = ck/\varepsilon$ and then the frequencies are only red-shifted: $\omega \Longrightarrow \omega/\varepsilon \simeq \omega(1 - 2GM/rc^2)$ because of the action of the static gravitational potential, but the dispersion relation is the same for the two polarization modes, like vacuum.

It is well known, in fact, that a static gravitational field can influence only direction and magnitude of the velocity of light, unlike the gravitational field of a rotating body, which can also produce a rotation of the plane of polarization (see for instance [8,9]).

In the presence of torsion, on the contrary, we can see from (4.18) that the usual degeneracy of the two polarization modes is destroyed ($\omega_+ \neq \omega_-$) and then torsion gives rise to a rotation of linear polarization even in a flat space-time.

Consider in fact a linear polarized wave sufficiently far from the central body where $\varepsilon \Longrightarrow 1$; in this case (4.18) is reduced to

$$\omega_{\pm}^2 = c^2k(k \mp \eta) \qquad (4.19)$$

and these are the dispersion relations in a flat apce-time with a constant torsionic background.

The electric field can be written as a linear combination of the two polarization modes putting

$$\bar{E}(t) = \bar{x}_1E_1 + \bar{x}_2E_2 = \mathrm{Re}(1/2)\{\bar{x}_1(E_+ + E_-) - i\bar{x}_2(E_+ + E_-) \qquad (4.20)$$

where $\bar{x}_1 \equiv (1,0,0)$, $\bar{x}_2 \equiv (0,1,0)$. We obtain from (4.17)

$$E(t) = A\cos\left[kx_3 - (1/2)(\omega_+ + \omega_-)t\right]\left[\bar{x}_1\cos(1/2)(\omega_+ - \omega_-)t - \bar{x}_2\sin(\omega_+ - \omega_-)t\right] \qquad (4.21)$$

and then if at $t_0 = 0$ the polarization vector is along the x_1 direction, after a time $t_1 = \pi/(\omega_+ - \omega_-)$ it will be in the x_2 direction. The plane of polarization rotates then with an angular velocity $\Omega = (\omega_+ - \omega_-)/2$, and developing (4.19) for wavelengths $\lambda = k^{-1}$ such that $\lambda\eta \ll 1$, we have, neglecting torsion contributions of the second order

$$\omega_{\pm} \simeq ck\left(1 \pm (1/2)\lambda\eta\right) \quad ; \quad |\Omega| = (c/2)\eta = 2\alpha G\hbar\rho_{\nu}/c^2 \qquad (4.22)$$

The plane of polarization of a wave propagation in a torsionic background, rotates then of an angle $\Delta\vartheta$ which is directly proportional to the length Δl of the path covered in this "medium"

$$|\Delta\vartheta| = |\Omega|\ \Delta t = (1/2)\eta\ \Delta l = 2\alpha(G\hbar/c^3)\rho_{\nu}\Delta l \qquad (4.23)$$

Even for very high neutrino densities [7], $\rho_\nu \sim 10^{39}\nu/cm^3$, however, one has a sensible rotation $\Delta\vartheta \sim 1$ only for $\Delta l \sim 10^{28}$cm ~ Hubble radius of the universe. This effect may become appreciable, therefore, only at a cosmological scale and/or if we consider the strong gravitational constant G_f instead of Newtonian constant G in (4.23) as for instance in the presence of superdense matter.

b) Dispersion

The second important effect induced by torsion is dispersion: the velocity of an e.m. wave in a space with torsion, depends on its frequency and polarization mode. From (4.18) we can deduce, in fact, that the phase velocity, $v = \omega/k$, is given

$$v_\pm = c\left(1 \mp (\eta/k)\right)^{1/2} = v(k) \qquad (4.24)$$

and that ω_+ becomes imaginary for $k < \eta$; therefore in this space a right-handed photon with a wave length greater than the critical value $\lambda_c = \eta^{-1} \sim (10^{67}/\rho_\nu)$cm, cannot propagate.

For $\rho_\nu \sim 10^{39}$, we find again a cutoff wave-length of the order of the radius of the universe $\lambda_c \sim 10^{28}$cm, but with the exception said before (as the case of superdense matter).

In general we can say that a space with curvature and torsion behaves like a dispersive medium described by an index of refraction n(k,x) such that

$$n^2\omega^2 = c^2k^2 \qquad (4.25)$$

from (4.18) we get

$$n_\pm^2 = \varepsilon^2\left(1 \pm (kc^2/\omega^2)\eta\right) = \varepsilon^2(1 \pm \lambda n_\pm^2\eta) \qquad (4.26)$$

from which

$$n_{\pm}^{2}(k,x) = \varepsilon^{2}/(1 \mp \varepsilon^{2}\lambda\eta) \qquad (4.27)$$

or, to the first order in G:

$$n_{\pm}^{2} = 1 + 4GM/rc^{2} \pm \lambda\eta \qquad (4.28)$$

In the absence of torsion ($\eta \Longrightarrow 0$) then $n_{+} = n_{-} = \varepsilon(r)$, i.e. the index of refraction is the same for the two polarization modes and it is only a function of the distance from the gravitating body. A light ray traveling at a distance R from its center is deflected by the well known angle δ

$$\delta = \varepsilon^{2}(R) - 1 = 4GM/Rc^{2} \qquad (4.29)$$

In the presence of torsion the gravitational index of refraction is altered by the factor $(1 \mp \lambda\eta\varepsilon^{2})^{-1}$ as shown by (4.27). The angle of deflaction δ', computed using trhe approximate expression (4.28) for the refractive index, turns out to be, to the first order in the torsionic contributions

$$\delta' = 4GM/Rc^{2}(1 \mp \lambda\eta) \qquad (4.30)$$

The joint action of a gravitational field and a torsionic background can induce, therefore, the spatial splitting Δx of the different wave-lengths contained in a wave-packet, on the plane normal to the propagation direction. The splitting is proportional to the distance l travelled through this medium

$$\Delta x = l\ \Delta\delta = \mp\ (4GM/Rc^{2})\eta\ l\ \Delta\lambda \qquad (4.31)$$

It is worth noticing, however, that according to the Heisenberg's uncertainty relations, the spatial resolution Δx is subject to the constraint $\Delta x\ \Delta k \geq 1$ i.e. $\Delta x \geq \Delta\lambda$.

In order that the effect predicted by (4.31) would be

measurable, it must be then

$$(4GM/Rc^2)\eta\, l > 1 \qquad (4.32)$$

Since in our universe $l \leq R_H$, where $R_H \sim 10^{28}$cm is the Hubble radius, and $2GM/Rc^2 \leq 1$, we have from (4.32) $2\eta\, R_H \geq 1$ i.e.

$$\rho_\nu \geq c^3/8\alpha G\hbar R_H \sim 10^{39}\mathrm{cm}^2 \qquad (4.33)$$

Therefore our previous estimate neutrino density [7] is the minimum allowable density which can give, on a cosmological scale a splitting compatible with the uncertainty pronciple $\Delta x \sim \Delta\lambda$.

If $\rho_\nu < 10^{39}$, then the cosmological neutrino sea is not able to produce, by means of torsion, any measurable diepersion effect, unless a spin-torsion coupling constant different from G is assumed [10] (as in the case of superdense matter or in the hypothesis of an energy dependent torsion coupling constant [11]).

c) photon mass

Consider again the effect of photon-torsion interaction and the eqs. (3.11) where appear a non linear term, in order to show that this non linear term can be interpreted as a mass term.

3. Non-linear term and massive electrodynamics

In general in a curved background, the electromagnetic wave equations assume the form:

$$\Box A^\mu + R^\mu_\nu\left(\left\{\ \right\}\right)A^\nu = 0 \qquad (4.34)$$

e.g. for a space of constant curvature $\Box A^\mu + \Lambda A^\mu = 0$, etc.

One can write down several possible curvature coupling such as $\Box A^{\mu} + R_{\mu\nu\rho\sigma}R^{\mu\nu\rho\sigma}A^{\mu}$, etc. In this case the equations are effectively equivalent to the wave equation with a photon mass term, i.e. are of the form

$$\Box A^{\mu} + (m_{\gamma}c/\hbar)^{2}A^{\mu} = 0 \qquad (4.35)$$

In the specific case we have here the coupling ofthe photon field to background torsion as given by eq.(3.11), the effective mass due to interaction with the torsion background of the universe is of the form:

$$(m_{\gamma}c/\hbar)^{2}A^{\mu} \cong (2\alpha/3\pi)\eta^{\mu\nu\rho\sigma}F_{\nu\rho}Q_{\sigma} \qquad (4.36)$$

We can roughly estimate the quantity on the right hand side of eq.(4.36) as follows: we can write for F_{0k} ,the dominant part of the field tensor,

$$F_{0k} = \frac{e}{R_{H}^{2}} \qquad (4.37)$$

R_{H} is the Hubble radius (we choose this radius because we are considering the average background torsion and e the electric charge (i.e.the virtual fermion in the torsion background of the universe). Also

$$A^{4} = e/R_{h}.$$

For a background spin density of $\bar{\sigma}$, the torsion vector can be written [12,13]:

$$\bar{Q} = \frac{4\pi G}{c^{3}}\bar{\sigma} \qquad (4.38)$$

$|\bar{\sigma}|$ was estimated in [14] and [15] and from a Machian viewpoint as:

$$|\bar{\sigma}| \approx (c^5/GH_o^2)\,(1/2\pi^2 R_H{}^3) \qquad (4.39)$$

H is the Hubble constant and R is the Hubble radius. This gives for $\bar{Q}$ from eq.(4.38), the magnitude $Q \approx 10^{-17}$cm for the torsion background. Then eq.(4.36) enables us to estimate for m_γ the effective mass of the photon as:

$$m_\gamma \approx 10^{-61}\text{g} \qquad (4.40)$$

which is consistent with the upper limit of $m_\gamma < 4\cdot 10^{-48}$g given from several experimental considerations [16,17].

As noted in already in [15], the existence of massive photons allows one to construct a superconductivity theory based on Einstein-Cartan theory. In fact we have a formal analogy of the torsion modified Maxwell equations with the London equation for superconductivity:

$$\nabla^2 B = (1/\lambda_L^2)B,$$

with λ_L the penetration depth giving an effective photon mass.

Here with torsion we have analogously $\nabla^2 H = (4\alpha/3\pi)H\cdot Q$ (for a zero background charge density) with the effective photon mass as estimated above due to interaction with torsion background.

In ref.[15] we had the analogy with symmetry breaking and we had obtained an expression for the boson mass, in terms of the critical magnetic field as:

$$m^2c^3/2eh = B_{crit.} \qquad (4.41)$$

(B was produced by the torsionic background, in accordance with $B \cong (8\pi/3c)(2\alpha G)^{1/2}\sigma$ (see ref.[12,15]); this is again analogous to the modification of B in a spinning superconductor as $B_{eff} = B + 2\mu_o n_s e\lambda_L^2\omega$ (with n_s the density of superfluid electrons); 2e is the charge of the pairs).

Using for m the value obtained above for photon mass, would give for e (i.e. the net excess charge, or charge on photon which is strictly zero by gauge invariance), the value $10^{-48}\mathbf{x}$ electron charge (q), which is more stringent compared with the earlier limit of $10^{-30}\mathbf{x}$ q from neutrino[18] and cosmic ray background [19].

Now for the universe B ~ 10^{-6}Gauss. Using for e the net excess charge constrained to below ~ 10^{-30}per electron charge (for arguments for this number from neutrino background see ref.[18] and from isotropy of cosmic rays ref.[19]), we get from eq.(4.41):

$$m_\gamma \approx 10^{-61}g \qquad (4.42)$$

which also agrees with the value (4.40) obtained above. That a small photon mass could be connected with a small violation of charge conservation is in the spirit of ref.[20]. However, in contrast to the London equation, we do not have a violation of gauge invariance.

So in conclusion a modification of the Maxwell equations due to the interaction with background torsion of the universe can give rise to an effective mass $\approx 10^{-61}$g, which is well within current observational limits.

In order to see in a more general way the problem of massive electrodynamics we would like to compare the problem of electrodynamics in Einstein-Cartan space-time, that is in space-time with torsion with the Chern-Simons gauge invariant massive electrodynamics. Some Physical consequences are outlined, however emphasizing the fact that in our case there is no arbitrary parameter, torsion Q playing the role of Chern-Simons parameter κ.

Whenever a photon mass is invoked in electrodynamics as has been done by several authors at different times [21-25] gauge invariance is necessarily broken as is well known. So whenever limits are put on a photon mass in different contexts [24,25], it is understood that gauge invariance and conformal invariance of Maxwell's equations is broken. As regards the conformal invariance of Maxwell's equations once we introduce a curved space-time the propagation of a photon is no longer conformally invariant (i.e. we have bending of light). Only in the special case of a conformally flat space are the Maxwell's equations conformally invariant. In general in a curved space-time, conformal invariance of Maxwell's equations is broken so that the curvature term can be interpreted as a mass term. For instance general covariance would modify the wave equation as:

$$\Box A^{\mu} + R^{\mu}{}_{\nu}A^{\nu} = 0 \qquad (4.43)$$

However it must be noted that this is not unique. One could write down as well $\Box A^{\mu} + R_{\alpha\nu}R^{\alpha\nu}A^{\nu} = 0$ or even $\Box A^{\mu} + R_{\gamma\nu\alpha\beta}R^{\gamma\nu\alpha\beta}A^{\mu} = 0$ etc. All these expressions would be generally covariant. The photon mass would be related or constrained by the overall curvature of space as $m_{\gamma}^{2} \propto R$.

The above considerations were for a curved space with a symmetrical connection. For a curved space with a nonsymmetrical connection it is also well known that minimal coupling does not preserve gauge invariance [26,27]. Suggested solutions involve invoking additional constraints on torsion or introduce other coupling prescription [28].

Classically one can of course retain gauge invariance by assuming that photons do not couple to the background torsion field. This complete decoupling could of course be valid only classically. The quantum description of the electromagmetic field would imply that a photon can for a small fraction of time disintegrate into a virtual fermion pairs and since massive fermions can minimally couple to torsion, this pair can interact with torsionic background inducing indirectly a photon-torsion coupling which can preserve the gauge invariance of the theory.

The Maxwell equations modified in this manner in the presence of torsion could be written as (see eqs.(3.11) of prevous chapter)

$$\begin{aligned} \partial_\nu F^{\mu\nu} &= 4\pi J^\mu + (2\alpha/3\pi)\eta^{\mu\nu\rho\sigma}F_{\nu\rho}Q_\sigma \\ \partial_{[\mu}F_{\sigma\nu]} &= 0 \end{aligned} \tag{4.44}$$

($\alpha = e^2/\hbar c$). These equations are not only compatible with current conservation $\partial_\mu J^\mu = 0$ (in the case of constant torsion i.e. $\partial_\nu Q_\sigma = 0$) but are <u>fully gauge invariant</u>.

In this modification of Maxwell's equation in the presence of background torsion we effectively have an 'extra' contribution to the sourse term on the right hand side of the first of eq.(4.44), something like a correction to the current densty J^{μ}. Also we note that the parameter Q_j which has the dimensionality of <u>inverse length</u> controls the torsion modification. Q can be defined through the spin density σ of the backgrounf matter as:

$$Q = 4\pi G\sigma/c^3. \qquad (4.45)$$

Q as defined in eq.(4.45) has the dimensions of inverse length. This additional term in eq.(4.44) can thus be pictured as providing a wavelength cut-off at large distances greater that Q^{-1} for the gauge invariant electric and magnetic fields. In the static case, we have a correction to the charge density.

We shall now point out some analogies between this additional gauge invariant term induced in the electrodynamic equations by torsion and the Chern-Simons modification of electrodynamics.

The gauge invariant modification of electrodynamics introduced by a torsion background as discussed above has a parallel with the Chern-Simons modification of electrodynamics which is currently receiving much attention [29]. With the Chern-Simons term we have the modified Maxwell equation given as:

$$\partial_{\mu}F^{\mu\nu} + (\kappa /2)\varepsilon^{\nu\alpha\beta}F_{\alpha\beta} = 4\pi J^{\nu} \qquad (4.46)$$

where the parameter κ has inverse length dimensionality and controls the Chern-simons additional term. This additional term gives rise again to a massive yet gauge invariant electrodynamics.

This is the usual interpretation of the Chern-Simons term which is a coupling between field and potential $A_\mu F^{\mu\nu}$, analogous to the coupling between torsion and field, i.e. here Q plays the role as the Chern-Simons parameter κ.

The Lagrangian density for Chern-Simons electrodynamics involves coupling between field $F^{\mu\nu}$ and potential A_μ thus:

$$\mathcal{L} = -(1/4)F_{\mu\nu}F^{\mu\nu} + (\kappa/4)\varepsilon^{\mu\nu\lambda}F_{\mu\nu}A_\lambda \tag{4.47}$$

The resulting field equations given by eq.(4.46) are invariant under

$$A_\mu \Longrightarrow A_\mu + (1/c)\partial_\mu\Lambda \tag{4.48}$$

The torsion modified part of the eletrodynamic action is given by [30]:

$$L = (\alpha/16\pi)\int d^4x\,\sqrt{-g}\;\varepsilon^{ijk}A_iF_{jk}Q \tag{4.49}$$

which is also invariant under the gauge transformation given by eq.(4.48).

Q is defined in terms of the background spin density through eq.(4.45) and being an inverse length is formally similar to the inverse length Chern-Simons parameter κ [31] (see also eq.(4.69)).

Notice also the field-potential coupling in eq.(4.49).

The fact that eqs.(4.47) and (4.49) correspond to a fully gauge invariant theory with massive photons can be seen by rewriting for instance eq.(4.46) using the dual fields given by

$$^{\bullet}F_{\nu} = (1/2)\varepsilon_{\nu\alpha\beta}F^{\alpha\beta}. \qquad (4.50)$$

Thus equation (4.46) becomes:

$$\left(g^{\mu\nu} + (1/\kappa)\varepsilon^{\mu\nu\alpha}\partial_{\alpha}\right)\,^{\bullet}F_{\nu} = 0 \qquad (4.51)$$

that we can write also in the form:

$$(\Box + \kappa^2)\,^{\bullet}F_{\beta} = 0 \qquad (4.52)$$

where the photon mass is related to κ as:

$$(m_{\gamma}c/\hbar)^2 = \kappa^2 \qquad (4.53)$$

In the torsion case, m_{γ}^2 is related to Q^2, with Q given by eq,(4.45).

So essentially what we have is a case of massive yet gauge invariant electrodynamics, 'massive' in the sense that we have a length parameter Q^{-1} entering as a cut-off (determined by torsion) which can be considered as equivalent to a mass term if one associate it with the corresponding Compton wave length, thus $m_{\gamma} \propto Q$. For vanishing torsion $Q = 0$, i.e. $Q^{-1} = \infty$, we recover the usual Maxwell electrodynamics with infinite field range and zero photon mass. We thus have the important consequence that we have no infrared divergence in electrodynamics as we have a wavelength cut-off at Q^{-1}. For instance on a cosmic scale Q arising from background torsion due to spin density of matter in the universe has a value of $Q \approx R_H^{-1}/\alpha$ giving the effective cut-off wavelength on a cosmic scale as $\approx R_H\alpha \approx 10^{26}$cm, R_H being the Hubble radius. This corresponds to an effective photon mass (from $\hbar/m_\gamma c = R_H\alpha$) of $m_\gamma \approx 10^{-62}$g. This mass for the photon is only an effective mass which is induced by the background cosmological torsion. It is not something intrinsic to the photon mass (like the W-boson mass for example). This is only an effective mass induced by its propagation through the background spin density, so that gauge invariance is preserved. Moreover the photon is not a gravitationally bound system (either by Newtonian or strong gravity). In this sense it is remarkable to talk of an effective photon mass on a cosmic scale.

The additional term in eq.(4.46) would give rise to a modified Gauss law as:

$$\nabla E - \kappa B = e\rho(r,t) \qquad (4.54)$$

where $\rho(r,t)$ is the charge number density. This has the consequence that any field configuration with charge $q = e\int dr\, \rho(r,t)$ also carries a magnetic flux $\Phi = \int dr\, B(r,t)$. As remarked earlier the additional term which is still gauge invariant can be regarded as contributing an 'extra' charge which would give rise to a background magnetic field (in turn related to Q). For a discussion of the torsion induced primordial magnetic field see [32].

The main effect of torsion modification consists of introducing a gauge invariant infrared cut-off Q^{-1} playing the role of a finite range analogous to the usual interpretation of Chern-Simons electrodynamics as massive gauge invariant theory with the range entering through κ^{-1}. We have also the corresponding divergence of the axial current as:

$$\partial^{\mu} j_{\mu 5} = e^{2} \varepsilon_{\mu\nu\lambda\sigma} F^{\mu\nu} F^{\lambda\sigma} \qquad (4.55)$$

For the torsion case, where the spin-torsion coupling arises from the axial-vector part of the torsion tensor [32,33] written as $Q = \partial_{\mu}\phi$ we have analogously (here we consider propagating torsion since we are considering field theory):

$$\partial Q \propto \Box\phi = (G\alpha/c^{5}) \varepsilon^{\mu\nu\rho\sigma} F_{\mu\nu} F_{\rho\sigma} \qquad (4.56)$$

The approach can be easily extended to the non-abelian case, the corresponding gauge groups being SU(2) isospin for the Chern-Simons case and SL(2,C) for the torsion case [15]. In this respect Chern-Simons approach is a particular case of torsion development.

It is also interesting to note that in the Chern-Simons case, κ^{-1} (i.e. the length parameter) is quantized in units of g^{2}/mc^{2}, where g is the gauge coupling i.e.

$$\kappa^{-1} = (g^{2}/mc^{2}) n \qquad (4.57)$$

where n is an integer. Similarly in the torsion case, if we associate torsion with defects in space-time topology [15] Q^{-1}

is quantized as:

$$Q^{-1} = n(\hbar G/c^3)^{1/2} \qquad (4.58)$$

or more generally $\int Q\, dA = n(\hbar G/c^3)^{1/2}$ where the integral is over a surface.

To see some physical consequences of a massive photon, we like to say that in a recent paper [35] Vigier discusses some cosmological consequences of a non-zero photon mass. However the photon mass in this paper is discussed in the sense of the usual Proca theory which is non gauge invariant unlike what we have here.

In our case we should also point out the analogy with the London equation which implies a modification of the electrodynamic equations inside a superconductor. For the London equation also we have a length parameter which gives the penetration depth which is the analogue of Q^{-1} here. For our case we have for instance, i.e. the torsion modified equation:

$$\nabla^2 H = (n\alpha/3\pi) H \cdot Q \qquad (4.59)$$

where the formal analogy with London's equation would again suggest Q^{-1} as the penetration depth, the photon again acquiring an effective mass, the important difference being that the London equation violates phase invariance wheras eq.(4.59) does not. Since torsion is also connected with magnetism [12],[15],[32] we can invoke the critical magnetic field given by:

$$B_{crit} = (m_\gamma^2 c^3/q\hbar) \qquad (4.60)$$

where q is an extra charge on the field as given by the additional term. This would also be consistent with the 'superfluid' background approach of Vigier [35].

Using for B_{crit} the primordial field of $\approx 10^{-7}$G and for $m_\gamma \approx 10^{-62}$g as estimated from the background torsion, we have a constraint on the extra charge as :

$$q \approx 10^{-48}e \qquad (4.61)$$

where e is the electric charge. This is much more stringent than the limits given by the isotropy of the cosmic ray background [36] which is $q \approx 10^{-32}e$. Thus this limit on the photon mass fixed by the background torsion, also constrains the photon charge to be $< 10^{-48}e$.

The well known dispersion relation for massive photons i.e. an effective refraction index $n \approx (1 + (m_\gamma{}^2 c^4/2\hbar^2 f^2)$, would for $m_\gamma \approx 10^{-62}$g imply a $(1 - v_\gamma^2/c^2) < 10^{-26} f^2$ and anisotropy changes of $\delta v_\gamma /c < 10^{-29}$, which are virtually unobservable.

Unlike the case of neutrinos which couple directly to torsion, photons do not couple directly to torsionic background [33]. In the case of neutrinos we had thus oscillations induced by torsion even for massless neutrinos [37]. We also had a magnetic moment for massless neutrinos induced by torsion [38].

It is very interesting to note that the above constraints on the torsion induced photon mass and charge also gives a photon magnetic moment of

$$\mu_\gamma = q\hbar/m_\gamma c \approx 10^{-20}\mu_\beta \qquad (4.62)$$

(μ_β = Bohr magneton). This although extremely small, could cause polarization changes in extremely strong magnetic fields like in neutron stars. If we have an effectively massive photon state propagating in a torsion background, we can have photon oscillations from one state of polarization to another. The oscillation distance would be given by (analogously to the case of neutrinos considered in [38]):

$$l_\gamma = E_\gamma / \Delta m_\gamma^2 c^3 \qquad (4.63)$$

where E_γ is the photon energy. The probability of oscillation in a propagation distance l is given by $\exp(-l_\gamma/l)$. For microwave background photons: $E_\gamma \approx 10^{-4}$ev. The lack of asymmetries in the microwave background would imlpy $\exp(-l_\gamma/l) \approx 10^{-5}$. If we take $l_\gamma \approx R_H$ to avoid inconsistencies, then eq.(21) implies $\Delta m_\gamma^2 < 10^{-52}$g, more stringent than current experimental limits.

4. Anyons and torsion

We now consider, in analogy to electromagnetic anyons as charge-magnetic flux composite, the gravitational anyons as mass-spin (energy-spin) composite. We will see that torsion interaction has many features in common with the Chern-Simons term so supporting the possibility of a gravitational anyon arising from torsion without introduction of Chern-Simons term (and so without any arbitrary parameter).

The concept of anyons is related to multiple connectedness of the two-dimensional rotation group. This comes out because, due to the abelian character of SO(2), spin is not quantized and thus can take any fractional value. For the same reason, we cannot understand the quantization of charge from just the U(1) gauge group of electromagnetism which is abelian. One would have to embed the U(1) group in a non-abelian gauge group of higher rank describing a unified interaction. In the framework of general relativity with torsion, the appropriate gauge group was shown to be the group SL(2,C) [15].

Now recently there has been a lot of interest in particles which can obey fractional statistics and occur in two-dimensional systems, the so called anyons [39]

Two-dimensional systems occur in various situations in condensed matter physics and it has been conjectured that anyons play an important role in understanding the fractional quantum Hall effect [40] and may be relevant to high temperature superconductivity [41].

When a wave function of a particle in two dimensions acquires under 2π rotation a phase different from ± 1 , we call the particle 'anyon' and when the wave function of two identical particles in two dimensional quantum mechanics acquires under exchange of the particles a phase different from ±1, the particles are called anyons, rather than bosons (+1) or fermions (-1). In three dimensions, the rotation group is non-abelian (i.e.SO(3)), and therefore spin is quantized, as there are commutation relations between the components, The existence of anyonic states in the presence of Chern-Simons term has been indicated [42]. In a recent paper [43] it was pointed out that the anyon can best be visualized as a charge-magnetic flux composite state, i.e. a particle with charge orbiting around a point source of magnetic flux carries a fractional angular momentum. But an analogous state was described for a particle with energy E orbiting a gravitational field generated by a spinning source, i.e. an energy-spin gravitational anyon state.

Now to see better how this energy-spin state is induced by torsion, we remember that in earlier papers [12], [34], it was pointed out that the torsion around a spinning particle can generate a magnetic moment given by

$$B = (1/c)(\alpha G)^{1/2}\sigma \tag{4.64}$$

(σ being the spin density). We also had the formal torsion, magnetic field analogy as

$$\overline{Q}\cdot\overline{S} \Rightarrow \overline{\mu}\cdot\overline{B} \tag{4.65}$$

Q being the torsion vector, arising from the Einstein-Cartan equations. The quantization of spin, led to the generation of torsion induced string defects in space-time topology, the string tension was shown to arise from spin-torsion interaction [44], [45].

This would also hold in two dimensions, with the exception that the spin can acquire any fractional value, the string carrying a corresponding magnetic flux; i.e. the world sheet of a torsion induced string defect lies on a closed surface Σ where then string carries a flux Φ. So we have a formal similarity with the Aharamov-Bohm (AB) interaction of a magnetic solenoid with an electrically charged particle.

The 3-form field strength can similarly be expressed as curl of a 2-form potential as torsion is described by a 3-form field strength in the gauge theoretic picture wherein it is possible to construct worm-hole solutions [15], [46], [47], where the torsion part of the Lagrangian was

$$\int d^4x \sqrt{-g} \; [(1/4x) S^{\mu\alpha\beta} S_{\mu\alpha\beta} \qquad (4.66)$$

Thus we have

$$H = dB, \quad \text{and} \quad Q = (1/2\pi)\int_{\Omega} H = (1/2\pi)\int_{\Sigma} B \qquad (4.67)$$

where Σ is the boundary of Ω.

The equivalent AB phase is

$$\exp(+i\int B) = \exp(2\pi i Q) \qquad (4.68)$$

This is the AB phase acquired by the defect which behaves like an anyon as the phase can take any fractional value. As torsion is related to dislocation density we can, in the dynamical picture, visualize the density fluctuations to propagate , as phonons (like in a solid). The corresponding change in torsion can be expressed as gradient of a scalar, i.e. $Q \Rightarrow \partial_{\mu}\phi$, so that effectively the anyon system here behaves like a phonon-screw dislocation, the screw helicity being related to the axial vector part of the torsion [47]. A well known property of the anyon system is the asymmetry with respect to parity. Here we know that the presence of torsion naturally implies a violation of parity.

The effective charge is the energy corresponding to time translations instead of the space translations. The deep reason for the existence of anyon states described here hence lies in the same topological three-dimensional gravity [43] but now in three space-like dimensions.

In application to superconductors there have been suggestion to test this experimentally by passing polarized light or particle beams through the superconducting copper oxide planar domains with so far inconclusive results. We had noted the formal similarity (see eq. (4.49)) between the Chern-Simons action given by

$$(\alpha/16\pi)\int d^4x\sqrt{-g}\ \ (\varepsilon^{ijk}A_iF_{jk})\cdot k \qquad (4.69)$$

and the torsion modified Maxwell equations:

$$(\alpha/16\pi)\int d^4x\sqrt{-g}\ (\varepsilon^{ijkl}A_iF_{jk}Q_l) \qquad (4.70)$$

with the torsion introducing a <u>scale</u> similar to the

Chern-Simons parameter (arbitrary) k.

The corresponding equations, in both cases (owing to the appearance of a scale in the theory), are strikingly similar to the London equations of superconductivity, which describe solutions containing flux tubes (i.e.lines of magnetic flux) with a penetration depth fixed by the scale (Q in the case of torsion and k for the Chern-Simons case). This magnetic flux cuts across the superconducting surface, so that a circulating charge acquires a traditional angular momentum (an anyon is often visualized as a two-dimensional system pierced by a thin flux tube which is just what we have here). So we have a way of describing anyons in a framework involving torsion.

There are other ways of understanding this. For instance in ref.[43] the gravitational anyon system was established by identifying (2 +1) dimensional electrodynamics coupled with gravitation as a (2 + 1) dimensional effective theory of the (3 + 1) dimensional Kaluza-Klein unification where the "fourth co-ordinate describes the internal space of the charge. However in ref.[34] we showed that by an appropriate gauge transformation the Einstein-Cartan Lagrangian can be reduced to an action in 4 dimensions involving curvature and spin which can be written as just curvature in 5-dimensions, i.e. the spin playing a more fundamental role than the charge by giving a particle magnetic moment generated through torsion. Thus here the internal co-ordinate describe the spin or magnetic moment, the periodicity of this co-ordinate giving fractional angular momentum. Thus the procedure of ref.[43] to generate an anyon in the Kaluza-Klein framework can be repeated in the present framework with torsion generating the magnetic moment (or flux) through the 'extra' co-ordinate whose periodicity generate fractional angular momentum as

$$l = n - e\Phi/2\pi \tag{4.71}$$

(n an integer).

That the energy-spin (or mass-spin) composite can form a gravitational anyon can also be considered from the torsion viewpoint. We give two different arguments. By analogy with the magnetic flux (i.e. $\int B\, dA$) we consider the vorticity flux $\int Q\, dA$ (Q is the torsion) (following the analogy between B and Q).

The vorticity quantum would be given by (analogous to the magnetic quantum flux):

$$\int Q\, dA = \hbar/m \tag{4.72}$$

Thus corresponding to the quantum relation between flux Φ and charge q as:

$$\Delta\Phi \cdot \Delta q = \hbar c \tag{4.73}$$

we have the analogous relation between the vorticity and mass as:

$$\Delta\Phi_m \cdot \Delta m = \hbar \tag{4.74}$$

($\Delta\Phi_m$ is in units of $\hbar/m$ just as $\Delta\Phi$ is in units $\hbar c/e$). In fact in liquid helium one observes this phenomenon of quantized vortices, just as in a superconductor one has quantized flux $\hbar c/e$.

This would imply that just as the charge-magnetic flux composite forms anyons in superconductors, the mass-spin composites (or mass-vorticity or energy-vorticity composites) must form anyon like systems in superfluid helium. As the torsion is connected to the axial vector, here again we would expect violation of parity. As phonons and rotons are know to propagate in superfluid helium, these systems would be the analogue of phonon-screw dislocations. They should polarize light or a particle beam passing through the liquid in a preferemtial direction which would be an indication of their presence. In superconductors the corresponding anyonic systems are formed by charge-magnetic flux composites. Here the multivaluedness of the wave function is traded for magnetic flux located at origin and electric charge located at particle such that $q\Phi/\hbar c = \vartheta$.

Torsion also trades magnetic flux for electric charge through spin by the same kind of relation.

It has been suggested [48] that anyons can undergo contact interactions by breaking of scale invariance. These contact interactions have a strength determined by the A-B magnetic flux. However we know that torsion introduces an additional term in the Dirac equation of the form:

$$\approx (3/8)\, Q_{\mu\nu\lambda}\, \varepsilon^{\mu\nu\lambda\rho}\, \gamma_{\rho 5} \qquad (4.75)$$

where $Q_{\mu\nu\lambda}$ is related to the spin tensor density $J_{\mu\nu\lambda}$. This term breaks the scale invariances, the scale being related to Q as noted above. So we indeed have contact interaction! In fact we do have the torsionic contact interaction between spinning particles, the axial vector part violating parity. This also happens in the Chern-Simons case. It should be possible to experimentally see whether forces between vortices in superfluid helium are different for parallel and antiparallel spins.

Anyons are supposed to be connected to the quantum Hall effect; the conductivity (ratio between current and magnetic field) being in fractional units of $\hbar c/e$. As we know that torsion generate a magnetic field, it would also imply a connection for the Hall effect involving the spin density in addition to the electric charge.(the conductivity now depending on both spin and charge). The coupling between spin and charge could give rise to coupled spin-charge density waves.

A gravitational anyon in ref.[42] was constructed by adding the Chern-Simons term to the Einstein action. The metric was shown to be a conical space-time as:

$$ds^2 = -[dt + \mu d\phi]^2 + d\rho^2 + \rho^2 d\phi^2 \qquad (4.76)$$

This describes a 'string-like' configuration, the defect angle δ being related to the string tension μ as $\delta \cong 2\pi(1 - 4\pi G\mu)$.

However as we saw in ref.[45], (see also Chap.XI) the string tension can arise from the spin-torsion interaction, the torsion interaction having many features in common with the Chern-Simons term. This again supports the possibility of a gravitational anyon arising from torsion <u>without</u> introducing a Chern-Simons term.

References

[1] V.de Sabbata and M.Gasperini - Lettere al Nuovo Cimento, 28,234.(1980). See also V.de Sabbata and M.Gasperini - Nukleonika 25, 1373 (1980)

[2] P.S.Letelier - Phys.Lett. 54A, 351 (1975)

[3] T.M.Davis and J.R.Ray - J.Math.Phys. 16, 75 (1975)

[4] J.Royer - Phys.Rev. 174, 1719 (1968)

[5] P.Bandhyopadhyay, P.r. Chauduri and S.K.Saha - Phys.Rev. D1, 377 (1970)

[6] L.Stodolski - Phys Rev.Lett.34, 110 (1975)

[7] V.de Sabbata and M.Gasperini - Lett.Nuovo Cimento 28, 181 (1980)

[8] J.Plebanski - Phys.Rev.118, 1396 (1960)

[9] Yu.S.Vladimirov and I.F.Iscakov - Izvestia Fizika 8, 45 (1970)

[10] V.de Sabbata and M.Gasperini - Lett.Nuovo Cimento 21, 328 (1978)

[11] V.de Sabbata and C. Sivaram - Astrophys.Space Sci. 158, 347 (1989)

[12] V.de Sabbata and M.Gasperini - Lett.Nuovo Cimento, 27,133 (1980).

[13] V.de Sabbata and M.Gasperini - In "The Origin and Evoluton of Galaxies" ed.by Venzo de Sabbata,World Sci.Singapore, (1981).pg.181

[14] V.de Sabbata and C.Sivaram - Nuovo Cimento,102B,107 (1988)

[15] V.de Sabbata and C.Sivaram - in "Modern Problems of Theoretical Physics (Festschrift for Professor D.Ivanenko) ed.by P.I.Pronin and Yu.N.Obukhov (World Sci.Singapore, (1991) p.143

16] L.Davis,Jr. A.S.Goldhaber and M.M.Nieto - Phys.Rev.Lett. 21, 1402(1975)

[17] G.Bicknell. - Gen.Phys.Math.,10, 407 (1977)

[18] C.Sivaram - Progr.Theor.Phys.,82,215 (1989).

[19] S.Orito and M.Yoshimura - Phys.Rev.Lett.54, 2457 (1985)

[20] P.A.Lyttleton and H.Bondi - Proc.Roy.Soc.London A252, 313 (1959)

[21] A.Einstein - Ann.Phys.18, 121 (1917)

[22] L.Bass and E.Schrödinger - Proc.of the Royal Society of London 232 (1955)

[23] L.Garcia de Andrade General Relativity and Gravitation, 64, 2969 (1990)

[24] T.L.Chow - Lettere al Nuovo Cimento,31, 289 (1981)

[25] A.Barnes and J.Sergle - Phys.Review Letters, 35,1117 (1975)

[26] F.W.Hehl, P.von der Heyde, G.D.Kerlich and J.M.Nester Review of Modern Physics,48,393 (1976)

[27] K.Hayashi and A.Sasaki - Nuovo Cimento B,45,205 (1978)

[28] M.Novello - Physics Letters, 59A, 105 (1976)

[29] R.Jackiw and S.V.P.So-Young Pi - Phys.Rev.Letters, 64, 2969 (1990)

[30] V.de Sabbata and M.Gasperini - Phys.Rev.D,23, 2116 (1981)

[31] V.de Sabbata and M.Gasperini - Lettere al Nuovo Cimento, 30, 503 (1981)

[32] V.de Sabbata and C.Sivaram - Nuovo Cimento A,100, 919 (1988)

[33] V.de Sabbata and M.Gasperini - Physics Letters,77A, 300 (1980)

[34] V.de Sabbata and C.Sivaram - Nuovo Cimento B,105,1181 (1990)

[35] J.P.Vigier et al. - Physics Letters 154A,203 (1991)

[36] C.Sivaram - Progress of Theoretical Physics,82,115 (1989)

[37] V.de Sabbata and M.Gasperini - Nuovo Cimento A,66, 479 (1981)

[38] V.de Sabbata and C.Sivaram - Nuovo Cimento B,105,603 (1990)

[39] F.Wilczek - Phys Rev.Lett. 48, 1144 (1982); 49, 957 (1982)

[40] Z.Zhang, T.Hansen and S.Kivelson - Phys.Rev.Lett. 62, 82 (1984)

[41] A.Fetter and R.Laughlin - Phys.Rev.B39, 9679 (1989)

[42] S.Deser - Phys.Rev.Lett. 64, 11 (1990)

[43] Y.M.Cho, D.H.Park, C.G.Han - Phys.Rev.D43, 1421 (1991)

[44] C.Sivaram and V.de Sabbata - Ann.der Phys.(Leipzig) 48, 362 (1991)

[45] V.de Sabbata and C.Sivaram - Ann.der Phys.(Leipzig) 48, 419 (1991)

[46] V.de Sabbata and C.Sivaram - Int.Journ.Theor.Phys. 30, 123 (1991)

[47] V.de Sabbata and C.Sivaram - in "Gravitation and Modern Cosmology, the Cosmological Constant problem" ed.by Antonino Zichichi, Venzo de Sabbata and Norma Sánchez, Plenum Publ. Corp.N.Y.Vol.56, 1992 pag.19

[48] G.R.Hagen - Phys.Rev.Lett.64, 503 (1990)

Chapter V

DIRAC EQUATION IN CURVED SPACE-TIME WITH TORSION

1. Introduction

Intrinsic spin became a compelling concept after the Einstein-de Haas [1] experiment (although wrongly interpreted initially by them in terms of orbital angular momentum!), further reinforced by advance in atomic spectroscopy leading to the formulation of the exclusion principle and finally culminating in its elegant incorporation into special relativity through the Dirac equation. It was subsequently recognized as a universal property of elementary particles and the unification of special relativity with quantum mechanics brought about by Dirac equation led to the prediction and understanding of a wide variety of new phenomena such as antiparticles, the connection between spin and statistics, the CPT theorem, automatic explanation of the electron's magnetic moment, etc. Neither special relativity nor quantum mechanics though both are complete theories in themselves could have individually predicted or accounted for such a wide variety of phenomena. It was only their unified description through the Dirac equation that led to these developments. The question then arises as to whether the incorporation of intrinsic spin into general relativity (a complete theory by itself) will lead to equally spectacular results. When formulating the theory of general relativity Einstein not being aware of the existence of intrinsic spin (despite the indications of the Einstein-de Haas [1] experiment which was done at around the same time!) naturally did not incorporate it into the structure of space time linking only the mass or energy momentum via the ten components of the metric tensor. Cartan [2] showed that the interaction of intrinsic spin with geometry must necessarily lead to a geometry with torsion present, the antisymmetric part of the connection now being connected to the intrinsic spin density.

In a pioneering paper in 1929, Fock and Ivanenko [3] while attempting to fit the Dirac electron into Einstein's general relativity, introduced the concept of local frames in the theory. The action of the Lorentz or Poincaré group on the local frame became the real prototype of a local gauge symmetry. We shall give a brief description of the steps leading to the Fock-Ivanenko coefficients. The so called vierbein or 16-component tetrad fields $t_a^\alpha(x)$ introduced by Weyl [4] bear to the metric tensor the same relation as Dirac's γ-matrices bear to the unit matrix, i.e. choose t_a^α ($a = 1,..4$) such that

$$t_a^\alpha \, t_b^\beta \, g_{\alpha\beta} = \delta_{ab} \, , \qquad \delta_{ab} = \begin{bmatrix} -1 & & & \\ & -1 & & \\ & & -1 & \\ & & & +1 \end{bmatrix} \tag{5.1}$$

Also $t_\alpha^a \, t_\beta^b \, \delta_{ab} = g_{\alpha\beta}$, t_α^a being the inverse matrix of t_a^α . If

L^b_a is a space-time dependent Lorentz matrix, then the transformation law for tretads can be written

$$t^\alpha_a = L^b_a L^\alpha_b \tag{5.2}$$

and
$$t = \det(t^\alpha_a) = \sqrt{-g} \tag{5.3}$$

We can thus interpret the Lorentz group as the group of rotations of the tetrad. If γ^α are the space-time dependent Dirac matrices, we have:

$$\gamma^\alpha = t^\alpha_a \gamma^a \tag{5.4}$$

where the γ^a are the usual constant Dirac matrices satisfying

$$(1/2)(\gamma^\alpha\gamma^\beta + \gamma^\beta\gamma^\alpha) = g^{\alpha\beta} \tag{5.5}$$

Consider now a spinor $\psi(x)$ which has the transformation law $\psi \Rightarrow S\psi$ (S is the space-time dependent spinor representation of the Lorentz matrix L given as $S^{-1}\gamma^a S = L^a_b \gamma^b$) when the tetrad is rotated as $t^\alpha_a = L^b_a t^\alpha_b$. While dealing with gravitational interaction of spinor fields the fundamental field is no longer the metric tensor but the vierbein fields t^α_a , the metric tensor being defined in terms of a quadratic product $t^a_\alpha t^b_\beta \delta_{ab} = g_{\alpha\beta}$ of these fields, i.e. in the quantum picture we may say that the expectation value of this product represents the metric.

An interesting feature is that we have taken an aspect of the reference frame (t^a_α) and treated it as a dynamical field so that the sharp distinction between the physics and the reference system used for describing it is no longer present!

This is a conceptual contribution of the Fock-Ivanenko treatment [3]. The Dirac matrices have themselves become fields and are no longer constant as in a flat space-time background. Equations (5.4) and (5.5) describe their space-time dependence and in fact they could be considered the fundamental fields with the metric tensor described through equation (5.5). To ensure that the derivative of a spinor transforms like a spinor, i.e. $\psi_{;\mu} \Rightarrow S\ \psi_{;\mu}$ we need to introduce a covariant derivative of the spinor $\psi_{;\mu} = \partial_\mu \psi - \Gamma_\mu \psi$. The 24-component gauge field $\Gamma_\mu = \Gamma^{ab}_\mu \gamma_a \gamma_b$ transforms as:

$$\Gamma_\mu \Longrightarrow S\,\Gamma_\mu S^{-1} + (\partial_\mu S)\ S^{-1} \qquad (5.6)$$

The spin coefficients defined as

$$\omega^\alpha_{a\lambda} = t^\alpha_{a;\lambda} = \partial_\lambda t^\alpha_a + \Gamma^\alpha_{\sigma\lambda} t^\sigma_a \qquad (5.7)$$

are skew-symmetric, i.e. $\omega^\alpha_{a\lambda} = -\ \omega^\alpha_{\lambda a}$, i.e. they are a generalized Christoffel like asymmetric connection. With this, we can then define the complete covariant derivative (Cartan derivative) as:

$$t^\alpha_{a;\lambda} = \partial_\lambda t^\alpha_a + \Gamma^\alpha_{\sigma\lambda}\ t^\sigma_a - \omega^b_{a\lambda} t^\alpha_b = 0 \qquad (5.8)$$

The vanishing of the Cartan derivative in eq.(5.8) serves to determine the spin coefficients in terms of the derivatives of the tetrad just as in Riemannian geometry the Ricci identity $g_{\mu\nu;\lambda} = 0$ leads to the identification of the affine connection with the Christoffel symbols. We can write $\omega^{a}_{b\lambda}\gamma^{b} - [\Gamma_{\lambda}, g^{a}] = 0$. Multiplying this by γ^{a} and using the Dirac matrix identity $\gamma_{a}[\gamma_{b}, \gamma_{c}]\gamma^{a} = 0$ enables the identification of the components

$$\Gamma_{\mu} = -(1/4)\omega_{\alpha\beta\mu}\gamma^{\alpha}\gamma^{\beta} \qquad (5.9)$$

as the Fock-Ivanenko coefficients [3]. This definition is unique up to the addition of a vector multiple of the unit matrix, i.e. the Γ_{μ} are arbitrary up to a gauge

$$\Gamma_{\mu} \Longrightarrow \Gamma_{\mu} + A_{\mu}I \qquad (5.10)$$

(where I is the identity matrix) and this arbitrariness enables introduction of the electromagnetic potentials! Owing to this gauge invariance, we can define a new derivative as:

$$\gamma'_{a;\lambda} = \gamma_{a,\lambda} + [\Gamma'_{\lambda}, \gamma_{a}(x)] \qquad (5.11)$$

where $\Gamma'_{\lambda} = \Gamma_{\lambda} + b_{\lambda}$. Under space-time transformations, fields $\gamma^{\alpha}(x)$ and Γ_{α} transform as usual contravariant and covariant quantities. The commutator of two covariant derivatives of a spinor is defined as:

$$\psi_{;\mu\nu} - \psi_{;\nu\mu} = \Gamma_{\mu\nu}\psi - Q^{\lambda}_{\mu\nu}\psi_{,\lambda} \qquad (5.12)$$

where

$$\Gamma_{\mu\nu} = \partial_\mu \Gamma_\nu - \partial_\nu \Gamma_\mu + [\Gamma_\nu , \Gamma_\mu] \qquad (5.13)$$

(contracted respect to tetrad, see eq.(5.8)) and

$$Q^\lambda_{\mu\nu} = \Gamma^\lambda_{\mu\nu} - \Gamma^\lambda_{\nu\mu} \qquad (5.14)$$

defines (see eq.(5.8)) the torsion tensor. $\Gamma_{\mu\nu}$, the curvature spinor, transforms under tetrad rotations as:

$$\Gamma_{\mu\nu} \Longrightarrow S\, \Gamma_{\mu\nu} S^{-1} \qquad (5.15)$$

the fields γ^α transforming as:

$$\gamma^\alpha \Longrightarrow S\, \gamma^\alpha S^{-1} \qquad (5.16)$$

This gives the relation between the curvature tensor and curvature spinor as

$$\Gamma_{\nu\lambda} = (1/4) R_{\mu\rho\nu\lambda} \gamma^\mu \gamma^\rho \qquad (5.17)$$

For a torsion free space we recover the usual expression. The connecting coefficients for the geometry are:

$$g_{\mu\nu;\lambda} = 0, \quad \Gamma^\lambda_{\mu\nu} = \left\{ {\lambda \atop \mu\nu} \right\} + (1/2) Q^\lambda_{\mu\nu} \text{ etc.} \qquad (5.18)$$

In terms of the spin coefficients the Riemann tensor is:

$$R^{\alpha\beta}{}_{\lambda\rho} = \omega^{\alpha\beta}{}_{\lambda,\rho} - \omega^{\alpha\beta}{}_{\rho,\lambda} - \omega^{\sigma\beta}{}_{\rho}\omega^{\alpha}{}_{\sigma\lambda} + \omega^{\sigma\beta}{}_{\lambda}\omega^{\alpha}{}_{\sigma\rho} \qquad (5.19)$$

Variation of the Lagrangian constructed from $R^{\alpha\beta}{}_{\lambda\rho}$ (analogous to the Einstein Lagrangian) with respect to tetrad components and spin coefficients leads to the Einstein-Cartan equations, i.e. the 16 equations

$$R_{\alpha\beta} - (1/2)g_{\alpha\beta}R + \Lambda\, g_{\alpha\beta} = -\chi\, T_{\alpha\beta} \qquad (5.20)$$

and the 24 field equations

$$Q^{\lambda}{}_{\mu\nu} + \delta^{\lambda}_{\mu}Q_{\nu} - \delta^{\lambda}_{\nu}Q_{\mu} = \chi\, S^{\lambda}_{\mu\nu} \qquad (5.21)$$

with the torsion tensor related to the density of an intrinsic spin source.

2. <u>The Heisenberg-Ivanenko equation</u>

The Lagrangian density for the curved space (with torsion) Dirac equation thus simply becomes:

$$\mathcal{L} = (1/2)i\sqrt{-g}\ (\bar{\psi}\,\gamma^{\mu}\psi_{;\mu} - \bar{\psi}_{;\mu}\gamma^{\mu}\psi) + (1/2)i\sqrt{-g}\ \Lambda^{1/2}\bar{\psi}\psi \qquad (5.22)$$

(where we have also included a cosmological term). The ';' is the spinor covariant derivative. It is easily seen that the equation $i\gamma^{\mu}\psi_{;\mu} + m\,\psi = 0$ is invariant under general coordinate transformation $\psi \Longrightarrow \psi'$, $\gamma^{\mu} \Longrightarrow (\partial x'^{\mu}/\partial x^{\nu})\gamma^{\nu}$, $\Gamma_{\mu} \Longrightarrow \partial x^{\nu}/\partial x'^{\mu})\Gamma_{\nu}$ as well as under tetrad rotations $\psi \Longrightarrow S\,\psi$, $\gamma^{\mu} \Longrightarrow S\,\gamma^{\mu}S^{-1}$, $\Gamma_{\mu} \Longrightarrow (S\,\Gamma_{\mu} + \partial_{\mu}S)S^{-1}$.

However the additional quantities in the covariant derivatives will generate extra self interaction terms in the curved space Dirac equation. Thus equation (5.22) becomes:

$$\mathcal{L} = (1/2)it[\bar{\psi}\, \gamma^{\mu}\psi_{,\mu} - \bar{\psi}_{,\mu}\gamma^{\mu}\psi + \Lambda^{1/2}\bar{\psi}\,\psi\, \{\Gamma_{\mu},\gamma^{\mu}\}\psi] \tag{5.23}$$

the curly brackets denoting anticommutation, this term containing spin coefficients. Eq.(5.23) can be considered a conventional Dirac Lagrangian with the interaction term $(1/8)it\varepsilon^{\nu\lambda\mu}\, \bar{\psi}\{\gamma_{\nu\lambda},\gamma_{\mu}\}\psi$. The spin density is

$$\Sigma^{\nu\lambda\mu} = (1/4)it\bar{\psi}\,\{\gamma^{\nu\lambda},\gamma^{\mu}\}\psi = -(1/2)\varepsilon^{\nu\lambda\mu\sigma}\bar{\psi}\gamma_{\sigma}\gamma_{5}\psi \tag{5.24}$$

The dual of $\Sigma^{\nu\lambda\mu}$ which is completely skew-symmetric is the axial vector

$$A_{\rho} = (1/2)\bar{\psi}\,\gamma_{\mu}\gamma_{5}\psi \tag{5.25}$$

giving the torsion tensor as

$$Q_{\mu\nu\rho} = -\,\chi\, A_{\mu\nu\rho} \tag{5.26}$$

Now if we denote covariant differentiation constructed from the Foch-Ivanenko coefficients by $\psi_{;\rho}$ then (5.23) can be rewritten as

$$\mathcal{L} = (1/2)it(\bar{\psi}\,\gamma^{\mu}\psi_{;\mu} - \bar{\psi}_{;\mu}\gamma^{\mu}\psi + \Lambda^{1/2}\bar{\psi}\psi) + (1/2)i\Omega^{\alpha\beta\lambda}\Sigma_{\alpha\beta\lambda} \tag{5.27}$$

The mass occurring in the Λ-term [5,6] can be reexpressed in terms of a spinor current: $\hbar c\Lambda^{-1}\gamma^{\mu}(\bar{\psi}\ \gamma_{\mu}\gamma_{5}\psi)\gamma_{5}$.

The Lagrangian density given by eq.(5.27) finally yelds the field equations:

$$\gamma^{\mu}\psi_{;\mu} + (1/8)Q_{\mu\nu\lambda}t\varepsilon^{\mu\nu\lambda\rho}\gamma_{\rho s}\psi + (1/\Lambda)\gamma^{\mu}(\bar{\psi}\gamma_{\mu}\gamma_{5}\psi)\gamma^{5}\psi = 0 \quad (5.28)$$

The second term in (5.28) can be rewritten using (5.25) and (5.26) as:

$$(1/8)Q_{\mu\nu\lambda}\ t\ \varepsilon^{\mu\nu\lambda\rho}\ \gamma_{\rho s} = (3/8)\chi\ \gamma^{\mu}(\bar{\psi}\ \gamma_{\mu}\gamma_{5}\psi)\gamma^{5} \quad (5.29)$$

The eq.(5.28) can be finally expressed in the term

$$\gamma^{\mu}\psi_{;\mu} + L_{o}^{2}\gamma^{\mu}(\bar{\psi}\ \gamma_{\mu}\gamma_{5}\psi)\gamma^{5}\psi = 0 \quad (5.30)$$

where L_o is a constant having the dimensions of length and given by

$$L_{o} = \left[(3\hbar G/8c^{3}) + (1/\Lambda)\right]^{1/2} \quad (5.31)$$

Equation (5.30) has a form precisely resembling the Heisenberg-Ivanenko non linear spinor equation [7,8] which forms the basis for a unified theory of particles.

Heisenberg's [7] unified field theory is however a special relativistic theory (gravitation is not incorporated into the theory). It starts from the Dirac special relativistic equation and assumes that only one fundamental field exists in nature, a spinor-isospinor field satisfying:

$$i\,\gamma^{\mu}\partial_{\mu}\psi \;+\; L_{o}^{2}\gamma^{5}\gamma^{\mu}\psi\;\bar{\psi}\;\gamma_{5}\gamma_{\mu}\psi \;=\; 0 \qquad (5.32)$$

The equation (5.32) is a non linear spinor equation so that the spinor field ψ is self-interacting. The various elementary particles are supposed to be different bound states of ψ, while their interactions are supposed to arise from the non linear interaction term in (5.32). A fundamental length $L_o \approx 2\cdot10^{-14}$cm is inbuilt into the theory and plays an important role. It is indeed quite remarkable that we have been able to obtain the form equation (5.32) (with the strong gravity constant we get the right length $L_o \approx 2\cdot10^{-14}$cm with $G = G_f = 6.7\cdot10^{30}$cgs in eqs.(5.30) and (5.31)) by the coupling of the massless Dirac field to the metric field through torsion. Thus the torsion background in the context of strong gravity leads naturally to the non linear spinor equation of Heisenberg and Ivanenko with an inbuilt fundamental length of $\approx 2\cdot10^{-14}$cm.

3. Fundamental interaction and torsion

If in the Lagrangian for the curved space Dirac equation we introduce a distinction between the left- and rigth-handed components as $\psi_{\pm} = (1/2)(1 \pm i\gamma_5)\psi$, we would obtain [9,10]:

$$\mathcal{L} = t[(1/2)\bar{\psi}\, i\, \gamma^{\alpha}(1 + \gamma_5)\psi_{,\alpha} \;-\; (1/2)\bar{\psi}_{,\alpha} i\, \gamma^{\alpha}(1 - i\gamma_5)\psi] \tag{5.33}$$

giving rise to terms of the type $-(\bar{\phi}_{\mu}\gamma^{\mu}\phi_{\nu})(\bar{\phi}_{\mu}\gamma^{\mu}\phi_{\nu})$, where $\bar{\psi}\,(1 - \gamma_5) = \phi$ and $(1 + \gamma_5)\psi = \bar{\phi}$. This is precisely what is obtained from the (V - A) theory of weak interactions. The fundamental length L_o would now be related to the beta decay length $\sim (G_F/\hbar c)^{1/2} \approx 6\cdot 10^{-17}$cm. G_F is the Fermi constant of beta decay It would perhaps be desirable in the context of high energy particle interactions to have an 'energy dependent' fundamental length since the effective G (for the torsion coupling) would be energy dependent [11,12] scaling as E^{-2} with energy.

Thus at the Planck scale ($E \sim 10^{19}$Gev), G becomes the Newtonian constant G_N and the fundamental length L_o is just the Planck length. At $E_{Gut} \sim 10^{15}$Gev, G would have the value of the Gut's coupling related to G_F as $G_{Gut} \sim G_F(M_W/M_G)^2$, M_W is the weak intermediate boson mass $\sim 10^2$Gev. At 10^2Gev, G would be the Fermi coupling, and L_o the beta decay length, and at ~ 1 Gev , the interaction would be strong, described by the strong gravity constant G_f, $G_f \sim G_N(M_{Pl}/M_f)^2 = G_F(M_w/M_f)^2$ etc. (M_f = f-meson mass) with a corresponding $L_o \sim 2\cdot 10^{-14}$cm, i.e. scale of the strong interaction (see Fig.1 of ref.[12]).

So we have L_o scaling as $\sim G^{1/2}$, i.e. as E^{-1}. This seems more appropriate rather than have a single value for the fundamental length. Moreover in the Lagrangian describing the generally covariant (curved space) Dirac equation: $\gamma_\mu(x)\left[(\partial/\partial x^\mu) + \Gamma'_\mu\right]\psi = 0$, where $\Gamma'_\mu = \Gamma_\mu + B_\mu$ (where B_μ is composed of Dirac matrices), the requirement of general covariance and gauge invariance would restrict the choice of B_μ to those of $B_\mu = V \pm A$, $B_\mu = P$ or $B_\mu = P + S$ where P, S refer to scalar and pseudoscalar. For weak interaction we had $V \pm A$. For the same equation to describe strong (Yukawa) interactions we must have pseudoscalar and scalar interactions which are also naturally accomodate. So the curved space Dirac equation would also have contact interactions of the scalar and pseudoscalar type which could provide a framework for describing eschange interactions of the Ivanenko-Tamm [13] form for dealing with strong interactions. As remarked earlier, the gauge invariance $\Gamma'_\mu \Longrightarrow \Gamma_\mu + A_\mu I$ in the Fock-Ivanenko coefficients enables the trivial introduction of the electromagnetic four potential in the gauge covariant derivative in the above equation (charge from torsion as in Chap.VI (see also ref.[14]). We thus see the possibility of a unified description of electroweak, strong and gravitational interactions in the above framework of the curved space Dirac equation with background torsion and with an energy dependent fundamental length L_0 scaling as $G^{1/2} \propto E^{-1}$.

That torsion in space-time can give rise to a fundamental

length can also be considered from another angle. This is to treat space time as a deformable medium with elastic properties. In the geometrical description of crystal dislocations or defects it is known that torsion plays the role of defect density in the limit of dislocations having a continuous distribution If we consider a small closed circuit and write $l^{\alpha} = \oint Q^{\alpha}_{\beta\gamma} \; \delta A^{\beta\gamma}$, where $dA^{\beta\gamma} = dx^{\beta} \wedge dx^{\gamma}$ is the area element enclosed by the loop and $Q^{\alpha}_{\beta\gamma} = \Gamma^{\alpha}_{[\beta\gamma]}$ is, as usual, the torsion associated with the connection $\Gamma^{\alpha}_{\beta\gamma}$. l^{α} represents the closure failure, i.e. torsion has an intrinsic geometric meaning: it represents the failure of the loop to close analogous to the crystal case, l^{α} having a dimension of length. In the above equation the torsion can be related to the intrinsic spin, i.e. to the fundamental unit of intrinsic spin $\hbar$, by postulating that the defect in space time topology should occur in multiples of the Planck length $(\hbar G/c^{3})^{1/2}$. l^{α} would be identified with the fundamental length. This is justified by the fact that we can find wormhole solutions with torsion with a size corresponding to the Planck length [15,16]. As G scales with energy as E^{-2}, l^{α} would scale as E^{-1}. At 100 Gev, it would be the beta decay length, at 10^{19} Gev the Planck length, at ~ 1 Gev the strong interaction scale and so on. So that the energy dependent torsion coupling gives an energy dependent defect length l^{α}. However the intrinsic spin would always be $\hbar$ as $G_N M_{Pl}^2 = G_F M_w^2 = G_f M_f^2 = \hbar c$. $\hbar$ and c are always the universal fundamental constants. c describes the propagation in space-time while $\hbar$ has to do with the intrinsic defect built into the torsion structure of space-time through $l^\alpha = \oint Q^\alpha_{\beta\gamma}\, dA^{\beta\gamma}$ i.e. $\hbar$ is related to quantized timelike vector with dimension of length related to energy dependent intrinsic geometrical structures with torsion like the Planck scale, the beta decay scale etc., which give to the fundamental interactions their characteristic properties. The dislocations in space-time induce spin momenta or torsion stresses. The above formalism also carries over into supergravity. We can use a 'vielbein' formalism instead of vierbein. If V^a denotes the vielbein, ω^{ab} the corresponding spin rotation coefficients, ψ^a the gravitino field, then the torsion is $Q^a = dV^a + \omega^{ab} V_b - (1/2)\bar{\psi}\, \gamma^a \psi$, the curvature $R^{ab} = d\omega^{ab} - \omega^a_c \omega^{cb}$, with the corresponding non-linear Dirac equations. It can also be modified to include superstring formalism like the super poincaré-virasaro [17] algebra.

The non linear term in the Heisenberg-Ivanenko equation with its inbuilt fundamental length can lead to divergence free self energies for particles (since we have a natural effective cut off) and also to non singular cosmological models. In fact as the non linear term is identified with that of torsion induced by interaction of spins with curvature of space-time, we have the possibility of the effective Hamiltonian density becoming non positive at sufficiently high densities [18] (see also Chapter IX). In general we had:

$$(1/V)\langle H\rangle = (1/V)E_o - (3\chi/16)\langle\bar{\psi}\ \gamma_n\gamma_5\psi\rangle\langle\bar{\psi}\ \gamma_n\gamma_5\psi\rangle \qquad (5.34)$$

the second term going as N^2/a^6 becoming large at small a, as, for instance, in the very early universe. This would give a minimum value for the radius of a collapsing universe suggesting the possibility of a bounce due to the effective density becoming negative.

4. <u>Machian origin of spin</u>

In the Machian view for the origin of local inertial mass, we consider the local mass to be generated by its gravitational interaction with all other masses of the universe. Thus if the local mass is m, then its energy mc^2 would be equal to the gravitational energy of its interaction with all other masses $\Sigma\ m = M_U$ in the universe (located over an average distance R). We can write

$$m\ c^2 \approx GM_U m/R \qquad (5.35)$$

This enables us to write the relation

$$G\ M_U/R\ c^2 \approx 1 \qquad (5.36)$$

as an essential condition for the Machian view to be sustained. Theories of gravitation such as the Brans-Dicke theory which incorporate the Machian picture also lead to a relation such as eq.(5.36). If the interactions of spins (i.e. through torsion) are included in a framework such as the Einstein-Cartan theory, then in addition to the GM_U/Rc^2 term in the metric potential we get the term G^2S_U/c^6R^4 [19], where S_U is the total intrinsic spin of the universe. If by analogy to the case of mass, we insist on a Machian origin of spin, we would get corresponding to eq. (36) for the total mass M, the relation for the torsion term [20]:

$$G^2S_U^2/c^6R^4 \quad \approx \quad 1 \qquad (5.37)$$

as a constraint on the total spin of the universe. We thus get:

$$S_U \approx c^3R_H^2/G \quad = \quad c^5/G\ H_o^2 \qquad (5.38)$$

where R_H is the Hubble radius (H_o is Hubble constant). This gives $S_U \approx 10^{93}$g cm^{2}s^{-1}. We can also write

$$S_U \approx M\ c\ R_H \qquad (5.39)$$

(c the velocity of expansion $\approx$ light velocity). For a typical elementary particle (say the proton):

$$S_p \approx m_p c\ r_p \qquad (5.40)$$

(m_p the proton mass, S_p its intrinsic spin, r_p its radius, speed of internal rotation assumed $\approx$ c).

Then we have:

$$S_U/S_p \cong (M_U/m_p)(R_H/r_p) \qquad (5.41)$$

From the Dirac LNH, $M_U/m_p \sim 10^{80}$, $R_H/r_p \sim 10^{40}$ we have

$$S_U/S_p \approx 10^{120} \qquad (5.42)$$

Since we know S_U from the Machian condition for total spin of the universe (as given by eq.(5.37)), we get:

$$S_p \approx 10^{93}/10^{120} \approx 10^{-27} g\ cm^2 s^{-1} \qquad (5.43)$$

We can thus identify S_p , the intrinsic spin of an elementary particle, with the Planck constant ħ. So we have deduced its value solely from cosmological parameters in the spirit of Machian viewpoint.

5. Torsion effects in atomic spectroscopy

The formal analogy between spin-torsion interaction (given by S·Q) with the interaction of a dipole moment in a magnetic field (given by $\mu\cdot B$) [21,22], would lead us to expect a splitting of atomic energy levels due to this interaction. This would be just like the hyperfine splitting of the ground state of the hydrogen atom i.e. the energy difference between the singlet and triplet due to the interaction of the proton magnetic moment with the magnetic field produced by the electron. This of course gives the famous 21 cm line (or ~ 1420 MHz frequency) in the case of hydrogen. The Hamiltonian in this case is given as:

$$H = \mu_p \int d^3x \, \rho(r) \, B(r) \tag{5.44}$$

with

$$B(r) = \nabla \times A(r) \tag{5.45}$$

where

$$A(r) = \frac{\mu \times r}{r^3} = \int d^3r' \, M(r') \times \nabla \left[\frac{1}{|r - r'|}\right] \tag{5.46}$$

and M(r) = dipole density. Eq.(5.46) can also be written as:

$$A(r) = -\mu_e \times \nabla \int d^3r' \frac{|\psi(r')|^2}{|r - r'|} \equiv -\mu_e \times \nabla \phi(r) \tag{5.47}$$

for an electron in state $\psi(r')$.

In the case of torsion, we know that Q(r) has the form of a dipole field also, i.e. $Q(r) = (G/c^2)S/r^3$, where S is the spin and moreover $Q(r) = \text{const.}\nabla \phi(r)$, i.e. Q(r) can be expressed as a derivative of a pseudoscalar field [23,24]. So directly analogous to the magnetic case, we can write the Hamiltonian for the spin-torsion interaction between the proton and electron in the hydrogen atom as (we assume that the torsion coupling has the same magnitude as the weak coupling constant, from the analogy of the spin-torsion interaction with the (V - A) form of the weak interaction explored above):

$$H = (8\pi/3)(\hbar/2)(\hbar/2)(4\pi G_F/c^2)\int d^3r\ |\psi(r)|^2 \qquad (5.48)$$

where with a spherically symmetric $|\psi(r)|^2$, we write

$$\nabla^2\phi = 4\pi\ |\psi(r)|^2 \qquad (5.49)$$

As usual with $|\psi(r)|^2$ slowly varying over dimension of proton, we can use the Fermi's formula for $|\psi(0)|^2$, i.e. replace $|\psi(r)|^2$ with $|\psi(0)|^2$ for the ground state. This gives finally:

$$H = (8\pi/3)(\hbar/2)(4\pi G_F/c^2)(\hbar/2\pi a_o n^3)\sigma_p\sigma_e \qquad (5.50)$$

where a is the Bohr radius ($a = \hbar^2/m\ ^2$) and

$$\sigma_p\sigma_e = \begin{cases} +\ 1 & \text{for triplet} \\ -\ 3 & \text{for singlet} \end{cases}$$

and the energy difference due to spin-torsion splitting of the ground state of the hydrogen atom becomes:

$$\Delta E = 32\pi G_F m_e^3 e^6/3\hbar^5 c^2 n^3 \simeq 3.5\cdot 10^3 \simeq 3.5\ \text{kHz} \quad (\text{for } n = 1) \qquad (5.51)$$

(n = principal quantum number). So the frequency is 3.5 kHz for H_2. Unlike the usual 21 cm (or ~ 1420 MHz) line, this radiation would not be detectable on the ground, owing to reflection by earth's ionosphere. However a very low frequency radio telescope has been suggested for deployment on the moon's dark side [25]. This would enable such radiation to be detected. As the weak interaction constant enters the formula, the probability of the transition would be reduced by a factor of $10^{-5}/10^{-2} \sim 10^{-3}$, giving a lifetime for the transition ~ 10^{16}sec. The intensity of the 21 cm line falling on the earth from all sources (i.e. from the whole sky) is known to be ~ 1 watt. Here it would be reduced by a factor of ~ 10^{3}. The flux falling on the moon would be ~ 500 ergs/sec. It may be detectable with the technology developed in the next century. Hydrogen compressed to very high densities at ultracold temperature might give a detectable flux on the laboratory. For solid H_2 (density ~ 10^{25} atoms/cm^3), the energy from a cubic meter would be ~ $3 \cdot 10^{-9}$ ergs/sec, at the above frequency. The transition could also be stimulated with a suitable tunable maser operating at very low frequencies.

We had considered the spin-torsion interaction energy splitting associated with only the principal quantum number n. In general there would be also splitting of levels with l, s, quantum numbers. Here the energy splitting would be generally given as:

$$\Delta E = \mu_s \cdot B = g_e Z \left[(k\ e^2)/2(mc^2)\right] L \cdot S/r^3 \qquad (5.52)$$

where $\langle L \cdot S \rangle = [j(j + 1) - l(l + 1) - s(s + 1)]\hbar^2/2$ (5.53)

and $\langle r^{-3} \rangle = (Z/a_o n)^3 [l(l + 1/2)(l + 1)]^{-1}$ (5.54)

k is the momentum vector.

The magnetic field associated with torsion is

$$B = (8\pi/3c)(2\alpha G_F)^{1/2} \hbar \langle r^{-3} \rangle \quad (5.55)$$

This would then give the energy splitting for a nucleus of charge Z, as (for quantum numbers n, l and s):

$$\Delta E = - g_e \frac{e\hbar}{2m_e c} s \frac{8\pi}{3c}(2\alpha G_F)^{1/2}\hbar\left(\frac{Z}{a_o n}\right)^3 [l(l \quad 1/2)(l + 1)]^{-1} \quad (5.56)$$

This work out to be ≈ same order of magnitude as in the case of hydrogen, i.e. a frequency of few kHz. We may expect the spin-torsion coupling to give energy splittings in the motion of a spinor particle analogous to the Landau levels in a magnetic field. For a scalar particle of mass m, it is known that the Landau levels are given by:

$$E_n^2 = k_3^2 c^2 + (2n + 1)\left[\frac{eBS}{m c}\right]^2 + m^2 c^4 \quad (5.57)$$

where k_3 is the momentum vector of the translatory motion in the direction perpendicular to the magnetic field.

For the spinor particle, it is known that the additional coupling associated with the spin, lowers the particle frequency, thus:

$$\omega_n^2 = (k_3^2c^2/\hbar^2) + (2n + 1)(eB/mc)^2 - \left[\frac{eBS}{mc\hbar}\right]^2 + m^2c^4/\hbar^2 \tag{5.58}$$

For torsion we would have the analogous formula:

$$\omega^2 = (k^2c^2/\hbar^2) + (2n + 1)(S\cdot Q/\hbar)^2 - \frac{2mc^2 S\cdot Q}{\hbar} + m^2c^4/\hbar^2 \tag{5.59}$$

The analogous of the gyrofrequency eB/mc would be $S\cdot Q/\hbar$ with the energy levels $(n + 1/2)(S\cdot Q/\hbar)$. The Landau levels are expected to be observable in the cyclotron line radiation seen in neutron stars (where the magnetic fields are very large). The torsion effects would also be expected to be significant in such a case.

References

[1] A.Einstein and W.de Haas - Proc.k.Ak.Amsterdam 18, 696 (1915); for a detailed description of this experiment see Abraham Pais "Subtle is the Lord..." Oxford University Press 1982

[2] H.E.Cartan - Compt.Rend. 174, 437, 593 (1922); Ann.Ecole Normale 41, 1 (1924)

[3] V.Fock and D.Ivanenko - Zeits.Phys.56, 330 (1929); Compt.Rend.188, 1470 (1929)

[4] H.Weyl - Zeits.Phys.56, 718 (1929); Proc.Nat.Acad.Sci. (USA) 15, 323 (1929)

[5] Y.Nambu and G.Jona-Lasinio - Phys.Rev.122, 345 (1961)

[6] C.Sivaram, K.P.Sinha and E.C.G.Sudarshan - Found.Phys. 6, 616 (1976)

[7] W.Heisenberg - "Unified Theory of Elementary Particles" Interscience, N.Y.1966

[8] D.Ivanenko and D.Sardanashvily - Physics Reports 94, 1 (1983)

[9] V.de Sabbata and M.Gasperini - "Introduction to Gravitation" World Sci.Singapore 1985

[10] V.de Sabbata and M.Gasperini - Nuovo Cimento 21, 328 (1978)

[11] V.de Sabbata and C.Sivaram - Astrophys.Space Sci.158, 347 (1989)

[12] V.de Sabbata and C.Sivaram - Astrophys.Space Sci.165, 51 (1990)

[13] D.Ivanenko - "Elementary Particles and Gauge Fields" ed.by D.Ivanenko Mir Publ.Moscow 1964

[14] V.de Sabbata and C.Sivaram - Ann.Phys.(Leipz.) 47(1990)

[15] V.de Sabbata and C.Sivaram - in "Quantum Mechanics in curved Space-time" ed.by J.Audretsch and Venzo de Sabbata, Plenum Press N.Y.1990 p.485

[16] V.de Sabbata and C.Sivaram - Int.Journ.Theor.Phys.30, 123, (1991)

[17] T.Piran and S.Weinberg - "Strings and Superstrings" World Sci.Singapore 1988

[18] V.de Sabbata and C.Sivaram - "Negative mass and Torsion" Theoret.Physics Moscow 1992, p.79

[19] A.Trautman - Nature 242, 7 (1973)

[20] D.W.Sciama - Mon.Not.Roy.Astron.Soc.113, 34 (1953)

[21] V.de Sabbata and M.Gasperini - Lett.Nuovo Cimento 27, 133 (1980)

[22] V.de Sabbata and M.Gasperini - in "The Origin and Evolution of Galaxies" World Sci.Singapore 1982

[23] V.de Sabbata and M.Gasperini - Lett.Nuovo Cimento 30, 193 (1981)

[24] V.de Sabbata and M.Gasperini - Physics Letters 83A, 115 (1981)

[25] J.O.Burns, N.Duric, G.J.Taylor and S.W.Johnson - Sci.Am. 262, 18 (1990).

CHAPTER VI

THE TORSION COUPLING CONSTANT AND THE PROBLEM OF THE COSMOLOGICAL CONSTANT

1. Energy dependence of coupling constant

First of all we like to show that torsion coupling constant (which, as we shall see, depends on energy as E^{-2}) can pass through the values of the coupling constants of the other interactions during the evolution of universe. An energy dependent torsion coupling constant helps in a natural way to understand the ratios of the coupling strengths of the different fundamental interactions.

It has been emphasized by many authors [1 - 5] that the general structure of metric-torsion theories allows parity violating interactions, i.e. the torsion of space-time might be responsible for parity violation in weak interactions. It has been recognized that the torsionic contact interaction between two Dirac particles has a formal analogy with the weak interaction Lagrangian with the usual Einstein-Cartan static term for the field and can be written as:

$$\begin{aligned} \mathcal{L} &= iV_\mu(\Psi)J^\mu(\psi') - A_\mu(\psi)J^{5\mu}(\psi') \\ &= (-3/16)\chi\ [J_\mu(\psi)J^\mu(\psi') + J^5{}_\mu(\psi)J^{5\mu}(\psi')\ ] \end{aligned} \tag{6.1}$$

$$\begin{aligned} &J^\mu = \psi\gamma^\mu\psi, \qquad J^{5\mu} = \psi\gamma^5\gamma^\mu\psi, \qquad V_\mu = Q_{\mu\alpha}{}^\alpha, \\ &A_\mu = (1/4)\varepsilon_{\mu\alpha\beta\gamma}Q^{\alpha\beta\gamma}, \ (Q_{\alpha\beta\gamma} \text{ is the torsion tensor}) \\ &V_\mu = i(3/16)\chi J_\mu, \quad A_\mu = (3/16)\chi J^5{}_\mu, \end{aligned} \tag{6.2}$$

This may be written in the standard (V - A) form if at least one of the two fermions is massless, i.e. described by a two-component spinor, i.e. for $\gamma^{\mu}{}_{,\mu}\psi' = 0$ and $(1 - \gamma^5)\psi' = 2\psi'$, we have

$$\begin{aligned} \mathfrak{L} &= -(3/32)\chi\psi'\gamma_{\mu}(1 - \gamma^5)\psi(J^{\mu} + J^{5\mu}) = \\ &= -(3/32)\chi\psi'\gamma_{\mu}(1 - \gamma^5)\psi'\psi\gamma^{\mu}(1 - \gamma^5)\psi \end{aligned} \tag{6.3}$$

Thus we have a torsionic interaction lagrangian which is formally identical to the weak interaction four fermion (V - A) lagrangian except for the value of the coupling constant. This raises the possibility that torsion may provide a geometrical model for weak interactions just like curvature does for gravitational forces. However the coupling constant in the above lagrangian (6.3) is

$$3\chi/32 \;=\; 32\pi K_g\hbar^2/4c^2 \;\approx\; 10^{-81}\mathrm{erg}\cdot\mathrm{cm}^3 \tag{6.4}$$

whereas the Fermi weak coupling constant for the four-fermion or (V - A) theory is

$$G_F/\sqrt{2} \approx 10^{-50}\mathrm{erg}\cdot\mathrm{cm}^3 \tag{6.5}$$

So in order to have a complete identification of torsionic and weak interactions we must postulate a spin-torsion coupling constant which is different from the mass-curvature coupling in Einstein's equations by a factor [2], [3], [6], of

$$\chi'/\chi = \sqrt{8}/9\ (G_F c^2/\pi K_g \hbar^2) \sim 10^{31} \qquad (6.6)$$

One has the possibility of giving arbitrary value to the torsionic coupling constant. There is no compelling a priori reason why it should have the same value as the gravitational constant.

In fact one can give some arguments to justify identifying the torsion coupling G_T to the Fermi G_F. For instance the similarity to the four-fermionic interaction (i.e. of the torsion interaction of two Dirac particles) i.e. with the action

$$\int d^4x (G_T/\sqrt{2})\psi_1 \cdot \psi_2 \cdot \psi_3 \cdot \psi_4 \qquad (6.7)$$

implies that G_T has the dimensionality of mass to a <u>negative</u> power. More precisely: the action is to be dimensionless; ψ being a spinor has dimensions $(\mathrm{mass})^{3/2}$ so that $(\psi)^4$ has dimensions $(\mathrm{mass})^6$, d^4x has dimensions of $(\mathrm{mass})^{-4}$, so that G_T must have dimension of $(\mathrm{mass})^{-2}$.

On the contrary, in the Hilbert lagrangian for gravity i.e.

$$\int d^4x\ (1/16\pi G)\sqrt{-g}\ R\ , \qquad (6.8)$$

the coupling $(1/16\pi G)$ has the dimension of $(\mathrm{mass})^2$; since $\sqrt{-g}$

d^4x has dimensions $(mass)^{-4}$, R the curvature scalar has dimensions $(mass)^2$, then for the action to be dimensionless $(1/16\pi G)$ must have dimensions of $(mass)^2$. This means that the dominant contributions to $1/16\pi G$ would arise from the higher mass states, the highest masses (i.e. mediating or virtual particles of highest mass) contributing the most. Thus for the Newtonian gravity coupling, the Planck mass states would chiefly determine the value $1/16\pi G_N$.

G_T on the other hand in eq.(6.7) having dimension of $(mass)^{-2}$, would receive dominant contributions from the lower mass states, i.e. from the intermediate weak bosons (W etc.). This would fix its value at the Fermi constant G_F, for intermediate boson masses $\approx$ 100 Gev. The whole hierarchy of heavier mass bosons up to Planck mass would not significantly contribute to G_T as their effect in determining its value would drop as $(mass)^{-2}$.

So if torsionic coupling constant receives its contributions from the intermediate weak bosons it could have the value $\approx G_F$.

At last we know that the spin-2 f-mesons mediate strong gravity. Their mass is $m_f \sim 1$ GeV, i.e. about a hundred times smaller that the intermediate weak bosons. Thus as the coupling G_T scales as $(mass)^{-2}$, we have the corresponding coupling constant for strong torsionic gravity as $\approx 10^4$ times $G_F \approx$ strong interaction coupling constant.

This suggests that the torsionic coupling constant should

be energy dependent and would scale with the corresponding mass M of the mediating bosons as :

$$G_T \propto 1/M^2 \qquad (6.11)$$

Thus $G_T \equiv G_F$, the Fermi constant, if $M \approx M_W$, the intermediate boson mass. For $M \approx M_{Planck} \approx 10^{19}$Gev, i.e. if the energy increases to such high values that particles of mass $\approx M_{Planck}$ mediate the interaction, then,

$$G_T \equiv G_{Newtonian} = G_F(M_W/M_{Planck})^2. \qquad (6.12)$$

thus explaining the large difference in weak and gravitational couplings. For $M_f \approx$ 1Gev, the f-meson mass, the coupling becomes large, i.e. $G_T \approx G_f$, which is the strong gravity constant. Thus

$$(G_f/G_F) \approx (M_W/M_f)^2,$$

or (6.13)

$$(G_f/G_{Newt}) = (M_{Planck}/M_f)^2 \approx 10^{38}$$

Eqs.(6.12) and (6.13) enable us to understand the ratios of the coupling strengths of strong, weak and gravitational interactions. Now Grand Unified Theories predict nucleon decay to occur through effective four-fermion interactions ~ $G_{GUT}(\psi)^4$, mediated by exchanges of superheavy gauge bosons, mass $M_x \approx 10^{15}$Gev. Therefore the corresponding coupling constant G_{GUT} by analogy with the expression for Fermi weak interaction ($g_w/8m_w^2 \approx G_F/\sqrt{2}$) can be written as:

$$G_{GUT}/\sqrt{2} = g_w^2/8M_x^2 \qquad (6.14)$$

The nucleon decay rate by this four-fermion interaction is:

$$\Gamma \propto 1/t_N \propto G_{GUT}^2 m_N^5 \propto m_N^5/M_x^4 \qquad (6.15)$$

where m_N is the nucleon mass.

For $$G_{GUT} \approx G_F(M_W/M_x)^2 \approx 2\cdot 10^{-75}\text{erg cm}^3, \qquad (6.16)$$

this gives $t_N \approx 10^{32}$yr.

At GUT energies, ($\approx M_x$), quantum gravity effects would not much alter proton decay rates, as contributions of bosons of $M > M_x$ to G_{GUT}, scale as $1/M^2$. So that $M \approx M_{Planck}$ mass bosons would change G_{GUT} coupling constant by one part in 10^8 only!

This suggests that the same torsionic coupling constant at $M_x \approx 10^{15}$Gev, i.e. $G_T \approx G_{GUT}$ can be responsible for nucleon decay.

It has been suggested that all forces are due to a spin-curvature coupling [7]. That is, in curved space with torsion there is a coupling of the spin tensor $S_{\mu\nu}$ and the curvature tensor $R_{\mu\nu\alpha\beta}$ to give a force $(1/2)R^{\mu\nu}{}_{\alpha\beta}S_{\mu\nu}x^\beta$. It has been shown [7] that all the forces that are derivable from gauge theories are of this form: i.e. they are due to a coupling of spin and curvature of the above form when the concepts of spin and curvature are suitably generalized. The question then as to whether the coupling constant of the spin-curvature interaction can describe the strengths of the basic interactions has been answered if one considers an energy-dependent torsion coupling constant. We have seen that eqs. (6.12) - (6.16) above do suggest that the coupling strengths of all the fundamental interactions can be accomodated in the energy-dependent torsion coupling constant G_T. Now it was suggested earlier [8,9,10] that in the early universe, during the hadron era (in connection with the Dirac two-metric theory) the gravity constant was appropriately the strong gravity one (i.e. G_f). So the above relations would suggest the evolution of the torsionic coupling constant as follows: during the Planck epoch it was identical with the Newtonian constant G_{Newt}. During the GUT's epoch it had the value of the GUT's coupling (G_{GUT} which was the same for strong, weak and e.m.), during the electroweak era it was G_F (Fermi constant) and during the hadron era it was G_f. It then reduced by expansion of the universe in accordance with Dirac relation $G \sim t^{-1}$ to its present value of G_{Newton} (see Fig.6.1).

If one does not believe in variation of G according to Dirac, then after hadron era there was a phase transition from G_f to Newtonian value G_{Newton} (see Fig.6.2)

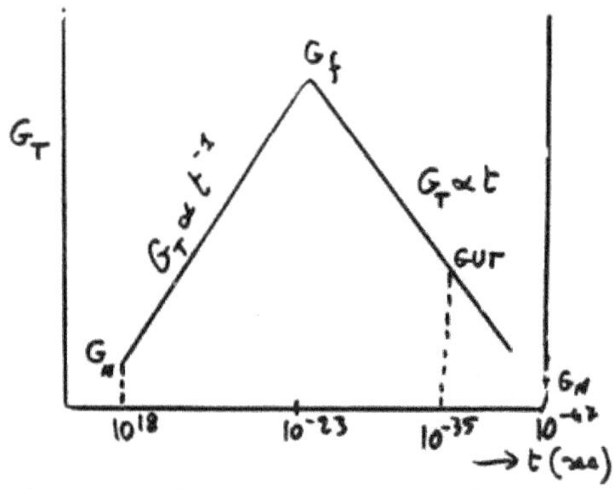

Fig.(6.1) Variation of torsionic coupling constant with time in the case that after hadronic era the gravitational constant follows Dirac hypothesis $G \sim t^{-1}$

One of the consequences of this energy dependent torsion coupling constant is that the spin contribution of matter sources can be large enough to cancel tha cosmological constant term at all stages in the early universe from Planck epoch.

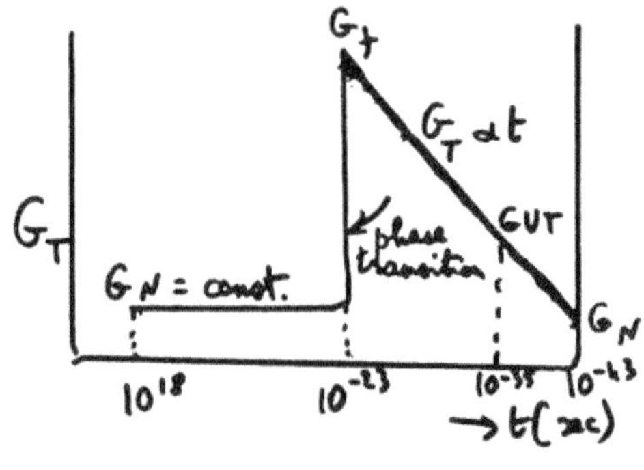

Fig.(6.2) Variation of torsionic coupling constant with time in the case that after hadronic era a phase transition from G_f to G_N has occured

2. Torsion and cosmological constant problem

The so called problem of the cosmological constant has been receiving considerable attention in recent years. A good current review of the situation concerning the problem is due to Weinberg [11]. Basically the problem is to explain why the effective cosmological constant is so small if not zero at the present epoch vastly less than the values one would expect from elementary particle physics theories. Anything that contributes to the energy density of the vacuum acts just like a cosmological constant the effective vacuum energy density being of the form

$$\rho_{\nu} \quad = \quad \Lambda_{eff} c^4/8\pi G \qquad (6.17)$$

There would have been large changes in the vacuum energy in the early universe as a result of phase transitions due to the breaking of some symmetry group as the expanding universe cooled. If the symmetry breaking takes place at some energy M, then the induced vacuum energy density is $\sim M^4$. For instance at the Planck epoch, $t \approx 10^{-43}$s., $M \approx 10^{19}$Gev, one would expect a large vacuum energy density term $\approx 10^{114}$ ergs cm^{-3}, corresponding to an effective cosmological constant $\Lambda_{Pl} \approx 10^{66} cm^{-2}$. This could arise as a result of scale invariance breaking and quantum gravitational contributions to the vacuum energy [12.13] in the very early universe at the Planck epoch. Again the GUTS phase transition at energies $\approx$

10^{15}Gev, would similarly induce another large Λ of $\Lambda_{GUTS} \approx 10^{50}cm^{-2}$. There would also be other large contributions from other symmetry breaking phase transitions at the electroweak scale for instance at somewhat later epochs in the early universe.

The question is what has happened to all these large contributions to the Λ-term. Why is the present value of Λ vanishingly small?

Weinberg summarize five different approaches undertaken in recent years to understand this question. He consider supersymmetry (exact global supersymmetry would indeed make the vacuum energy and hence Λ vanish). But we know that supersymmetry must be broken quite strongly and this would give a large contribution to Λ which would not vanish. There is no symmetry principle known (like gauge invariance in electromagnetism implying zero photon mass) which would make Λ vani sh exactly and Weinberg states that it is very hard to see how any property of supergravity or superstring theory could make the effective cosmological constant sufficiently small. He also resorts to the anthropic principle to explain why Λ is so small but this is a rather weak argument [14,15].Again most attempts have involved some sort of adjustment mechanism [16,17] which requires some extra scalar field, which evolves and acts as a counterterm to cancel the cosmological term. However it turns out that in all such attempts the scalar field must have some very special ad hoc properties and involves a lot of 'fine-tuning' at all stages, apart from there being no evidence of such extra fields. Other attempts have dealt with changing the structure of Einstein's equations but this also leads to several consistency problems. Recently there has been a lot of excitement about a new mechanism suggested by Coleman [18] which follows up an earlier work of Hawking which described how in quantum cosmology there could arise a distribution of values for the effective cosmological constant with an enormous peak at $\Lambda_{eff} = 0$. Coleman considers the effect of topological fixtures known as wormholes, consisting of two asymptotically flat spaces joined together at a 3-surface, and shows that the probability distribution or expectation value has an infinite peak at $\Lambda_{eff} \to 0$. However several objections have been raised, including the reality of wormhole existence, the use of Euclidean quantum cosmology (it is essential that the path integral be given by a stationary point of the Euclideanized action) which may have nothing to do with the real world. Moreover if the path integral has a phase, that might eliminate the peak in the probability distribution at zero cosmological constant. In short there are too many controversies with Coleman's very speculative proposal. One promising possibility which has not been considered so far in understanding the cosmological constant problem is the use of torsion in a framework such as the Einstein-Cartan (E-C) theory which is natural in considering the gravitational contributions of particles with spin which is indeed a universal property of elementary particles. In fact at sufficiently early epochs the energy content of the universe can indeed be spin dominated and the temporal evolution of the spin-density tensor is important in describing the cosmological dynamics [19], [20], [21].

In Einstein-Cartan theory the Lagrangian is the usual scalar curvature:

$$L_{E-C} = (-g)^{1/2} R(\Gamma) \qquad (6.18)$$

where Γ is non-symmetric affine connection:

$$\Gamma_{\alpha\beta}{}^{\mu} = \left\{ {\mu \atop \alpha\beta} \right\} - K_{\alpha\beta}{}^{\mu} \qquad (6.19)$$

and $K_{\alpha\beta}{}^{\mu}$ is the contorsion tensor which is related to the torsion tensor $Q_{\alpha\beta}{}^{\mu} \equiv \Gamma_{[\alpha\beta]}{}^{\mu}$ by

$$K_{\alpha\beta}{}^{\mu} = - Q_{\alpha\beta}{}^{\mu} - Q^{\mu}{}_{\alpha\beta} + Q_{\beta}{}^{\mu}{}_{\alpha} \qquad (6.20)$$

The dynamical spin density tensor can be written:

$$\tau^{\alpha\beta\mu} \qquad 1/(-g)^{1/2}\left(\delta L_m/\delta K_{\mu\beta\alpha}\right) \qquad (6.21)$$

where $\tau^{\alpha\beta\mu}$ is connected with torsion tensor by $T^{\alpha\beta\mu}$ = $\chi\tau^{\alpha\beta\mu}$ being $T^{\alpha\beta\mu}$ the modified torsion tensor:

$$T_{\alpha\beta}{}^{\mu} = Q_{\alpha\beta}{}^{\mu} + \delta^{\mu}_{\alpha} Q_{\beta\nu}{}^{\nu} - \delta^{\mu}_{\beta} Q_{\alpha\nu}{}^{\nu} \qquad (6.22)$$

The torsion algebra is related to the matter spin density and one can substitute spin for torsion everywhere. The variation with respect to $g_{\alpha\beta}$ and $K_{\alpha\beta\mu}$ gives the field equations:

$$G^{\alpha\beta}\Big(\{\ \}\Big) = \chi\,(T^{\alpha\beta} + \tau^{\alpha\beta}) \tag{6.23}$$

where $T^{\alpha\beta}$ is the usual energy momentum tensor and $\tau^{\alpha\beta}$can be considered as representing the contribution of an effective spin-spin interaction [22] i.e. product terms:

$$\begin{aligned}\tau^{\alpha\beta} = \chi\Big[&- 4\tau^{\alpha\mu}{}_{..[\nu}\tau^{\beta\nu}{}_{..\mu]} - 2\tau^{\alpha\mu\nu}\tau^{\beta}{}_{.\mu\nu} + \tau^{\mu\nu\alpha}\tau_{\mu\nu}{}^{\beta} + \\ &+ (1/2)g^{\alpha\beta}(4\tau_{\mu}{}^{\nu}{}_{.[\rho}\tau^{\mu\rho}{}_{..\nu]} + \tau^{\mu\nu\rho}\tau_{\mu\nu\rho})\Big]\end{aligned} \tag{6.24}$$

This is equivalent to an effective cosmological term. L_m , the Lagrangian for the matter fields, describes spinning fluid coupled minimally to the metric and torsion of Riemann-Cartan manifold.

The dynamical spin tensor can be written

$$\tau^{\alpha\beta\mu} = (1/2)S^{\alpha\beta}u^{\mu} \tag{6.25}$$

where $S^{\alpha\beta}$ is spin density and u^{μ} the 4-velocity of the fluid and $S_{\alpha\beta}u^{\beta} = 0$ holds. $S_{\alpha\beta}$ is associated with the quantum mechanical spins of elementary particles and effective sources of the gravitational field described by space-time averaging of $T^{\alpha\beta} + \tau^{\alpha\beta}$. Even if the spins are randomly oriented, the average of the spin-squared terms is not zero in general [23]

For unpolarized spinning fluid $\tau^{\alpha\beta}$ can be decomposed in two parts: the usual canonical energy-momentum tensor for spinning particles

$$\langle\tau^{\alpha\beta}\rangle = (1/2)\chi\, \sigma^2 u^\alpha u^\beta + (1/4)\chi\, \sigma^2 g^{\alpha\beta} \qquad (6.26)$$

where $\sigma^2 = (1/2)\langle S_{\alpha\beta} S^{\alpha\beta}\rangle$, and also the quadratic part corresponding to the spin-spin and spin-torsion term taht is:

$$\langle\tau^{\alpha\beta}\rangle = -\chi\, \sigma^2 u^\alpha u^\beta \qquad (6.27)$$

As regards $T^{\alpha\beta}$ we have

$$\langle T^{\alpha\beta}\rangle = (\rho + p) u^\alpha u^\beta - p\, g^{\alpha\beta} \qquad (6.28)$$

Of course we are considering only the case of unpolarized spinning fluid (so there are no antisymmetric terms), so we deal only with averaged squared terms: we do not require alignement of spins [23]. For alignment of spins we require a very large initial magnetic field, which owing to exponential expansion would be diluted by exp[2Ht] and regenerated in hadron era [24], [21] with strong gravity.

The simplest E-C generalization of standard big bang cosmology is obtained by considering universe filled with unpolarized spinning fluid and solving modified Einstein equations $G^{\alpha\beta}(\{\}) = \chi\, \vartheta^{\alpha\beta}$, where

$$\vartheta^{\alpha\beta} = \langle T^{\alpha\beta}\rangle + \langle\tau^{\alpha\beta}\rangle = (\rho + p - (1/2)\chi\,\sigma^2)\,u^\alpha u^\beta - (\rho - (1/4)\chi\,\sigma^2)g^{\alpha\beta} \qquad (6.29)$$

where ρ.p, and σ depend only on time. In the comoving frame $u^\mu = (0,0,0,1)$, we get the following modified field equations of the Robertson-Walker universe, which in general for $k \neq 0$ and $\Lambda \neq 0$ is of the form:

$$\dot{R}^2/R^2 = (8\pi G/3)[\rho - (2/3)\pi\, G\,\sigma^2/c^4] + \Lambda c^2/3 - kc^2/R^2 \qquad (6.30)$$

We immediately notice that the torsion term in eq. (6.30) (the second term within brackets) is of <u>opposite</u> sign to that of the cosmological constant term. This raises the possibility that a sufficiently large spin-torsion term in the early universe might cancel a correspondingly large cosmological constant. We shall see that this is indeed the case. For instance consider the universe at the Planck epoch when as we noted earlier the Λ term was $\approx 10^{66}cm^{-2}$imlying $\Lambda_{Pl}c^2 \approx 10^{87}$ in eq.(30). At $t_{Pl} \approx 10^{-43}s$, the universe had a density of $c^5/G^2\hbar \approx \rho_{Pl} \approx 10^{93}g\ cm^{-3}$, and as the particle masses were $\approx 10^{19}Gev \approx 10^{-5}g$, the particle number density was $n_{Pl} \approx 10^{98}cm^{-3}$, so that σ the spin density was

$$\sigma_{Pl} \approx 10^{98}\cdot 10^{-27}(\text{i.e. } n_{Pl}\cdot\hbar) \approx 10^{71} \quad (^{*}). \qquad (6.31)$$

This gives for the term $-(8\pi G/3)(2/3)\pi G\ \sigma_{Pl}^2/c^4$ (i.e.the torsion term in eq.(6.30) the value of $\approx -10^{87}$, which is exactly equal and of opposite sign to that of the cosmological term $\Lambda_{Pl}c^2 \sim +10^{87}$ so that the two terms would have cancelled each other in the early universe at the Planck epoch. We can see that they would continue to cancel at later epochs in the early universe.

The Λ term would evolve wlth temperature T as $\Lambda \propto T^2$ [25] being the spin density (i.e. proportional to number density of spins) would scale as T^3 ($n \propto T^3$ for relativistic particles). So σ^2 would scale as T^6. Now in an earlier paper [26] it was argued that the spin-torsion coupling in the early universe was energy dependent and scaled as $G \propto T^{-2}$, i.e. $G \propto T^{-2} \propto t$. So $G^2 \propto T^{-4}$, so that the $G^2\sigma^2$ term would scale as

---

(*) Note that, as we will show in Chp.XII where we introduce the concept of maximal force (i.e. $F_{max} = c^4/G$), we can have a maximal magnetic field given by

$$F_{max}/e \cong B_{max} = c^4/Ge \approx 10^{58}\text{Gauss} \qquad (6.32)$$

We notice that this is just the field we have at the Planck epoch (which agrees with what we have in eq.(2.23) of Chp.II). Now using the relation between spin density σ and magnetic field B, i.e. $B = (8\pi/3c)(2\alpha G)^{1/2}\,\sigma$, we have

$$\sigma_{max} = (3cB_{max}/8\pi)(2\alpha G)^{-1/2} \qquad (63.3)$$

Substituting for B the value of eq.(6.32) we get

$$\sigma_{max} \approx 10^{71} \qquad (63.4)$$

which is just what we have in eq. (6.31)

T^2, the same dependence on T as the Λ-term so that if they cancel each other at the Planck epoch, they woud also cancel at later epochs in the early universe.

So in short we have a more natural mechanism for understanding of a vanishing Λ term by the simple incorporation of spin effects (a universal property of particles!) in general relativity. As regards the possible behaviour of the cosmological constant after the hadron era we can say that at the Planck epoch, the temperature was $\approx 10^{32}$ K $\approx m_{Pl}c^2/k_B$, at the hadron era $T_h \simeq m_p c^2/k_B \simeq 10^{13}$ K and so on. Could one possibly have a residual cosmological constant? Since the universe could have a net angular momentum of $\approx 10^{94}$ [27], [28], [29] the second term in eq.(6.30) would imply a negative cosmological constant corresponding to a density $\approx G\sigma^2/c^4 \approx 10^{-30}$g cm^{-3} not far from the critical density. Until the hadron era we have the relation $G^2\sigma^2 \propto T^2$ so that this temperature dependence ensures cancellation of the Λ-term induced at the various phase transitions. After the hadron era the second term in eq.(6.30) continues to evolve with time since the universe has a net spin density, and the present value of $G\sigma^2/c^4$ is, as we have seen, $\approx 10^{-30}$g cm^{-3}. So also the second term in eq. (6.30) has a very rapid fall with time.

As far as the cosmological constant is concerned the present value would be expected to be comparable to the present curvature which is $\approx 10^{-56}$cm^{-2}. Since Λ scales as T^2 with temperature in the early universe, we would expect Λ_h in the hadron era, to be $\Lambda_{Pl}(T_h/T_{Pl})^2 \approx 10^{66}\cdot 10^{-40}$ (since $T_h \approx 10^{13}$ K, $T_{Pl} \approx 10^{33}$ K) that is $\Lambda_h \approx 10^{26}$ cm^{-2}. This corresponds to vacuum energy densities of 10^{94}g cm^{-3} and 10^{17}g cm^{-3} in hadron era, and agrees well with what is obtained in the strong gravity picture for hadron era. After hadron era the evolution of the cosmological constant with time would be expected to scale as t^{-2} [25]. So since $t_h \approx 10^{-23}$s, and since the present Hubble age is $\approx 10^{18}$s,, this would imply a present value of the cosmological constant of $10^{26}(10^{-23}/10^{18})^2 \approx 10^{-56}cm^{-2}$.

References

[1] F.W.Hehl and B.K.Datta - J.Math.Phys.12, 1334 (1971)
[2] C.Sivaram and K.Shina - Lett.Nuovo Cimento 13, 357 (1975)
[3] F.Kaempffer - Gen.Rel.Grav. 7, 327 (1976)
[4] V.de Sabbata and M.Gasperini - Gen.Rel.Grav.10, 731 (1979)
[5] V.de Sabbata and M.Gasperini - in "Spin, Torsion, Rotation and Supergravity" ed. by Peter G.Bergmann and Venzo de Sabbata, Plenum Publ.Corp. New York 1980 pg.139
[6] V.de Sabbata and M.Gasperini - Il Nuovo Cimento 21, 328 (1978)
[7] E.G.Harris - Il Nuovo Cimento 55B, 205 (1980)
[8] C.Sivaram and K.Sinha - Phys.Reports 51, 111 (1979)
[9] V.de Sabbata and P.Rizzati - Lett.Nuovo Cimento 20, 525 (1979)
[10] V.de Sabbata and M.Gasperini - in "Origin and Evolution of Galaxies" ed.by Venzo de Sabbata, World Sci.Singapore 1982 pg.181

[11] S.Weinberg - Rev.Mod.Phys.61, 1 (1989)

[12] C.Sivaram - Astrophys.Space Sci.125, 189 (1986) and 127, 133 (1986)

[13] C.Sivaram - Int.Jour.Theor.Phys. 25, 825 (1986)

[14] V.de Sabbata - Journal of the Tensor Society of India 1, 39 (1983)

[15] V.de Sabbata 1984] - Giornale di Astronomia - pg.275

[16] A.D.Dolgov - in "The very early universe" ed. by G.Gibbons et al. CUP Cambridge, 1982 pg.449

[17] F.Wilczeck - in "Proceed.of Erice 1983 Conference" ed.by A.Zichichi (Plenum, N.Y.) pg.208

[18] S.Coleman - Nucl.Phys. B307, 867 (1988)

[19] A.Trautman - Nature, 242, 7 (1973)

[20] V.de Sabbata - in "Gravitational Measurements, Fundamental Metrology and Constants" Kluwer Acad. Publ.The Netherlands 1988 pg.115

[21] V.de Sabbata and C.Sivaram - Il Nuovo Cimento 102B, 107 (1988)

[22] F.W.Hehl, P.von der Heyde, G.D.Kerlick and J.M.Nester - Rev.Mod.Phys. 48, 393 (1976)

[23] B.Kuchowicz: - Acta Cosmological 415, Z.3, 109 (1975)

[24] V.de Sabbata and M.Gasperini - Lett.Nuovo Cimento 27, 133 (1980);

[25] C.Sivaram, K.Sinha and E.C.G.Sudarshan - Found.Phys. 6, 717 (1976)

[26] V.de Sabbata and C.Sivaram - Astrophys.and Space Sci. 158. 347 (1989)

[27] V.de Sabbata and M.Gasperini - Lett Nuovo Cimento 25,489 (1979)

[28] V.de Sabbata and M.Gasperini - J.T.S.C., Journal of Tensor Society of India,Lucknow 1, 39 (1983)

[29] V.de Sabbata and M.Gasperini - Lett.Nuovo Cimento 37, 605 (1983)

CHAPTER VII

SPIN, MAGNETIC MOMENT AND ELECTRIC CHARGE

1. Is spin more fundamental than electric charge?

A classical point charge electron possesses an infinite amount of electrostatic self energy, owing to the divergence of $W = (1/8\pi) \int E^2 dV = (1/8\pi) \int (e^2/r^4) 4\pi r^2 dr = \int (e^2/2r^2) dr \Longrightarrow \infty$ as $r \Longrightarrow 0$. This is the so called infinite Lorentz divergence of the electron energy [1] Moreover this would lead to an unstable collapse of the point charge electron, there being no force to counteract the repulsive electrical self energy. Attempts to postulate internal stresses and pressures analogous to that of a fluid to counteract the electric energy were made by Lorentz, Poincaré and others but without much success . Born and Infeld [2], [3] gave a non linear electromagnetic model but this too had many inconsistencies. Again Casimir [4] invoked the zero point energy of the vacuum fluctuations of the electromagnetic field to compensate for the electrostatic self energy but did not quite succeed. In fact as far as the electron is concerned, the vacuum energy ~ $\hbar c/r_c$ (where r_c is the electron Compton length, $r_c = \hbar/m_e c$), is equal to the classical self energy ~ e^2/r_o, where r_o is the classical radius $e^2/m_e c^2$ since $r_o \sim (e^2/\hbar c) r_c \sim \alpha r_c$). One can argue that the charge is localized onto a region of radius r_o whereas the vacuum energy which is a quantum effect occurs around r_c. The tendency of electron to 'explode' under its repulsive self energy is compensated by the attractive quantum force due to vacuum fluctuations around r_o which exerts an opposing pressure of magnitude ~ $\hbar c/r_c^4$. But it is not clear from this that the electron would be stabilized around r_o. We shall consider the possibility that the spin of electron also plays a dominant role, the spin energy density leading to torsion could give rise to a stable classical electron.

Now the energy density of the electric field surrounding the charge is

$$\rho = E^2/8\pi = (1/8\pi)\frac{e^2}{r^4}, \qquad (7.1)$$

which when integrated over a volume, diverges for a point electron. The electron classical radius is obtained by just applying a cutoff at some radius r_0 and equating the resulting electric energy to the electron rest energy. For Born-Infeld type theories [3] the radius is $\approx e^2/E_o$, where E_o is some maximal field strength. For $r < r_o$, the electron field must deviate radically from that of a classical point particle and other effects must come in to play. The classical infinity becomes softened if we consider a non-linear model [5] due to the self energy, i.e. have a non linear Poisson equation of the type:

$$\nabla^2\Phi = -\frac{1}{2c^2}(\nabla\Phi)^2, \qquad (7.2)$$

whose radial part can be written:

$$\frac{1}{r^2}\frac{d}{dr}\left(r^2\frac{d\Phi}{dr}\right) = -a\left(\frac{d\Phi}{dr}\right)^2. \qquad (7.3)$$

This can be easily shown to have a solution:

$$\Phi = (1/a)\ \ln\,[1 - a/(kr)] + b \qquad (7.4)$$

which with a boundary condition $\Phi \Longrightarrow 0,\ r \Longrightarrow \infty$, becomes:

$$\Phi = (1/a)\ \ln\,[1 - a/(kr)] \qquad (7.5)$$

giving rise to a logarithmic divergence in self energy for a classical point particle. Quantum effects to the electron self energy are also known to diverge logarithmically, i.e. one has the usual formula:

$$\delta m/m \approx (e^2/4\pi)\ \ln\,[\hbar c/r]\ , \qquad (7.6)$$

and if one applies a cutoff at the Schwarzschild radius and insists on $\delta m/m \sim 1$, one has an empirical relation between the gravitational and electromagnetic constants, of the type $\alpha^{-1} \sim \ln\,(\hbar c/GM^2)$. Here we obtain logarithmic divergence in the classical case itself. In general one consider the source term in the Poisson equation to be made up of a bare mass term $m_o\delta(r)$ and electric charge term $e^2/8\pi c^2r^4$ and ρ, the net energy density, would be the sum of these. Then the non-linear Poisson equation for electrodynamics would be:

$$\nabla^2\phi = 4\pi G\rho - \frac{1}{2c^2}(\nabla\Phi)^2. \tag{7.7}$$

With $\Psi = e^{\Phi/2c^2}$, this becomes

$$\nabla^2\Psi = (2\pi G/c^2)\ \rho\ \Psi \tag{7.8}$$

Using $u = 1/r$ and with the boundary condition $\Psi \Longrightarrow 1$, $r \Longrightarrow \infty$, the solution of equation (7.8), is easily shown to be:

$$\Psi(r) = -\frac{\cosh\ (K - e/2r)}{\cosh\ K}, \tag{7.9}$$

or

$$\Phi(r) = -\frac{e\tanh\ (K)}{r} \tag{7.10}$$

enabling $M = e\tanh\ (k)$ to be defined as the mass of the system.

One can then easily show that the gravitational energy density almost compensates the electromagnetic self energy, i.e.:

$$\rho_{em} = e^2/8\pi r^4$$

$$\rho_{grav} = -(\nabla\Phi)^2/8\pi = -(1/8\pi)\ \mathrm{sech}\ (K - \frac{e}{2r})\frac{e^2}{r^4}.$$

$\rho_{total} = \rho_{em} + \rho_{grav}$ and integrating $\int_0^\infty \rho_{tot} dV$ gives e tanh (K) which is just the mass of the system.

Now we must also incorporate the electron spin which would modify Poisson's equation as [6]

$$\nabla^2 \Phi = 4\pi G_f (\rho - K_f \sigma^2) \qquad (7.11)$$

where ρ is ρ_{tot} which includes e.m. and gravitational energy density; $K_f = 8\pi G_f / c^4$, where σ is the spin density given by $S = (4/3)\pi\sigma R^2$, where S is the net spin of the system. The radial equation of motion has an exact solution.

$$R(t) = \left(\frac{3G_f s^2}{2mc^4} + 9\frac{G_f mt^2}{2} \right)^{1/3} \qquad (7.12)$$

(this is only interior solution because at the exterior we have electrostatic field) giving a minimum radius:

$$r_o = \left(\frac{3G_f \hbar^2}{8mc^4} \right)^{1/3} \sim 1.4 \cdot 10^{-13} \text{cm} \qquad (7.13)$$

Substituting for the electron mass $m_e \approx 10^{-27}$g and the strong gravity constant $G_f \approx 6.6 \cdot 10^{30}$cgs units, gives $r_o \approx 1.4 \cdot 10^{-13}$cm. This means that if we put t = 0 (in order to have (7.13)) we have the static configuration and this configuration is not point like but has a minimum radius.

If this is to be identified with the classical electron

radius, (that is $r_0 \Longrightarrow e^2/2mc^2$) we would get a relation for the electric charge as:

$$e = (3G_f \hbar^2 m_e^2 c^2)^{1/6} \sim 5 \cdot 10^{-10} \text{esu} \qquad (7.14)$$

We can now extend this model to strong interactions.

2. The proton, muon and other particles

In fact the proposed finite classical model for the electron with torsion and strong gravity can be extended to include not only the proton, but also the muon and other particles.

So we extend this model to strong interactions, the balance between strong gravity and torsion now giving rise to a finite proton radius obtained by simply replacing m_e in eq.(7.14) by the proton mass m_p. Indeed the equation (7.14) is basic in the sense that replacing m_e with the proton mass m_p, we get a finite classical radius for the proton as $r_p \approx 1.2 \cdot 10^{-14}$cm, here again the balance is between strong gravity and torsion i.e.

$$r_p = (3G_f \hbar^2/8m_p c^4)^{1/3} \approx 1.2 \cdot 10^{-14} \text{cm} \qquad (7.15)$$

At this point we like to stress (we will see better in the following sections) that the electric charge is a derived concept, with the spin $\hbar/2$ of the particle playing a more fundamental role entering through the torsion term and providing a finite classical radius by balancing the strong gravitational force between the constituents.

If we identify r_p now with the strong interaction radius $\approx g^2/2m_pc^2$ (g^2 is the strong coupling charge squared), then we have the relation:

$$g^2/e^2 = (m_p/m_e)^{2/3} \qquad (7.16)$$

which with $e^2/\hbar c \approx 1/137$ would give for the strong coupling $g^2/\hbar c \approx 1$, i.e. the strength of the strong interaction. Thus with just torsion entering through the spin of the particle, we are able to obtain both electromegnetic and strong coupling constants through G_f and the particle masses m_p or m_e. We can also connect all this with what we have said as regards the dependence of torsion coupling constant from the energy (going as E^{-2} with energy) in the early universe (see (6.11),(6.12),(6.13) in Chapter VI). In this case we had the relation

$$G_f\, m_p^2 = G_N\, m_{Pl}^2 = G_w\, m_w^2 \qquad (7.17)$$

where $G_w = G_F(c/\hbar)^2$, m_{Pl}, m_w are the Planck and weak intermediate boson masses, G_F, G_N are the Fermi and Newtonian constants. Thus for masses (energies) of the Planck mass, the coupling becomes G_N, the Newtonian value. Then eq.(7.15) gives the radius

$$\left(\frac{3G_N \hbar^2}{8m_{Pl}c^4}\right)^{1/3} \approx 1.5\cdot 10^{-33}\text{cm} \qquad (7.18)$$

which is the Planck length for particle of Planck mass, and for the intermediate boson mass, we get the beta decay length $\sim 6\cdot 10^{-17}$cm giving the corresponding radius for the boson.

The proton is thus supported by torsion at the finite radius r_p. So we have a stable proton which does not decay .

Eq.(7.13) would imply that the electron mass is such that it is in equilibrium at the radius given by (7.13) as a result of a balance between torsion and the strong gravielectromagnetic field. A heavier particle not undergoing strong interactions would not be supported by torsion at the above finite radius (of course this is not the case for a particle that, like the proton with strong charge 'g', undergoes strong interaction). So a system for which torsion is not sufficient to give rise to an equilibrium radius, must collapse. So if the electron were to undergo an energetic interaction leading to the absorption of a large amount of energy, with the net mass $M \gg m_e$, as the spin remains the same, torsion would not be able to support it at the radius r_0 and it would collapse. Then we have to consider the total mass (or energy) of the system with just mass and charge with negligible torsion thus giving (at any radius r) the balance between charge and strong gravity as:

$$M_{tot} = M_0 + e^2/2rc^2 - G_f M_{tot}^2/2rc^2 \qquad (7.19)$$

where M_0 is the "bare" mass of the particle, M_{tot} being the total net observed mass, which acts as the source for the strong gravitational field. Eq.(7.19) is readily solved for M_{tot} giving:

$$M_{tot} = - rc^2/G_f + \left[r^2c^4/G_f^2 + e^2/G_f + 2m_0rc^2/G_f\right]^{1/2} \qquad (7.20)$$

We do not consider the case of negative mass in this context. It is remarkable that eqs. (7.19) and (7.20) are rigoursly valid even in the context of general relativity, i.e. one obtains identical result given by eqs.(7.19) and (7.20) even if the full general relativity is considered, not just Newtonian gravity , as the last term in (7.19) would suggest.

Now since the system not supported by torsion must tend to collapse, we must take the $r \Longrightarrow 0$ limit of eq.(7.20). Thus (7.20) gives for the case when the particle collapses to a radius $r << r_0$, the limiting value for the mass, as:

$$M_{tot} = e/\sqrt{G_f} \qquad (7.21)$$

With the usual value for $G_f = 6.7 \cdot 10^{30}$ cgs units, the mass $e/\sqrt{G_f}$ has the value $\approx 1.9 \cdot 10^{-25}$g $= 106$ Mev which is just the muon mass! Thus the muon is not stable, since its mass is too large for it to be supported by torsion. It would decay into its stable counterpart, i.e. electron, which is supported by torsion. So torsion must play a role in its decay. But we know (see (6.7) of the previous Chapter) that the torsionic interaction between Dirac particles, i.e. torsionic Lagrangian is formally identical to the weak interaction (V - A) Lagrangian with the action given by:

$$\int d^4x \ (G_T/\sqrt{2}) \ \bar{\psi}_1\psi_2\bar{\psi}_3\psi_4 \qquad (7.22)$$

with the torsion coupling now being identified with the Fermi constant G_F (see also section 5). Again since the size of the system 'r' which has collapsed is << r_0 it would be of the order of the beta decay length when the energies are of the order of the intermediate boson mass mediating the weak interaction.

In the case of the intermediate boson (which takes part in only electroweak interactions) we have the analogous expression for the mass: $m_z = e/\sqrt{G_w}$ where G_w is related to the Fermi constant as $G_w = G_F(c/\hbar)^2$. So the appropriate coupling is G_w (in fact we have $G_w m_w^2 = \hbar c = G_f m_f^2$ etc.).

Notice that electroweak unification occurs when $G_w m_z^2 = e^2$ corresponding to a mass of $m_z \approx e/\sqrt{G_w}$ ($\approx$ 70 Gev)), while m_w ($\approx$ $(\hbar/G_w c)^{1/2}$ $\approx$ 250 Gev) represents the intermediate boson corresponding to the Fermi beta decay length. This particle can perhaps be found in the Tev accelerators now under constructions or being or being planned like the eloisatron. Considering the fundamental nature of e and G, we would expect to see a spectrum of masses corresponding to $e/\sqrt{G_N}$, $e/\sqrt{G_f}$,

$e/\sqrt{G_w}$ that is

$$G_N m_\alpha^2 = G_w m_z^2 = G_f m_\mu^2 = e^2 \qquad (7.23)$$

(with $m_\alpha = m_{Pl}\sqrt{\alpha}$). These masses correspond to balance betweeen electromagnetism and Newtonian gravity, electromagmetism and weak interaction, electromagnetism and strong gravity respectively. As regards the relations:

$$G_N m_{Pl}^2 = G_w m_w^2 = G_f m_f^2 = \hbar c \qquad (7.24)$$

this corresponds to the fact that at these energies (m_{Pl} for gravity, m_f for strong interaction and m_w for weak), all these interactions have the same coupling $\hbar c$. So we seem to have two sets of particle masses (7.23), (7.24) differing from a factor $\sqrt{\alpha}$ .Since the torsion-coupling constant G_T in the early universe is energy dependent (see Chp.VI and ref.[7] i.e. goes as E^{-2}, one would expect to have a spectrum of masses corresponding to $e/\sqrt{G_T}$ and $(\hbar c/G_T)^{1/2}$, where G_T can take the values G_N , G_{GUT} , G_w , G_f , which correspond (for various $e/\sqrt{G_T}$) to muon ($e/\sqrt{G_f} \simeq$ 105 Mev), to electroweak boson ($e/\sqrt{G_w}$ $\simeq$ 70 Gev) and perhaps to other fundamental particles of masses $e/\sqrt{G_N}$ ($\simeq 10^{17}$Gev), $e/\sqrt{G_{GUT}}$ ($\simeq 10^{13}$Gev) etc. Also corresponding to $(\hbar c/G_T)^{1/2}$, we have the Planck mass $(\hbar c/G_N)^{1/2}$ ($\simeq 10^{16}$Gev), the proton $(\hbar c/G_f)^{1/2}$ ($\simeq$1Gev), the weak boson $(\hbar c/G_w)^{1/2}$ ($\simeq$250Gev), and $(\hbar c/G_{GUT})^{1/2}$ ($\simeq 10^{14}$Gev) etc. Since the interaction is now identical to the weak (V - A) interaction, (see eq.(7.22)) the muon lifetime would simply be given by:

$$t_{\mu} = \left[7\ m_{\mu}^{5}c^{4}G_{F}^{2}/7680\ \pi^{3}h^{7}\right]^{-1} \simeq 2\cdot 10^{-6}s \qquad (7.25)$$

where we have the typical $G_F^2 m_\mu^5$ scaling for beta decay. We shall not consider the τ-lepton here, but its lifetime would be expected to be much shorter as it scales inversely as m^5 for the same coupling. It would be of order $\approx 10^{-12}$s (because $m_\tau/m_\mu \geq 10$ and $t_\tau/t_\mu \propto (m_\mu/m_\tau)^5 \approx 10^{-5}$) but since it is heavier than the proton, there may be some strong interaction effects to complicate the situation.

For strongly interacting particles, the lifetime would simply be ~ $\hbar/Mc^2$ from the uncertainty principle. Equivalently in the strong gravity formalism, using $G_f \approx \hbar c/M^2$, the lifetime would have the typical $G_f^2 M^3$ behaviour (for hadrons). Now consider again that charge is a derived concept from torsion. In this case what is more fundamental is the flux given by $\hbar c/e$ and in fact we have the uncertainty relation between flux Φ and charge q as:

$$\Delta\Phi\cdot\Delta q \geq \hbar c \qquad (7.26)$$

Since we obtained the particle magnetic moment and radius (and hence the flux) from <u>spin</u> through torsion, the charge Δq would be quantized from eq.(7.26) in units of e. So the question arises as to whether we can consider mass as also obtainable from spin through torsion. This becomes possible if by analogy with electric flux, we consider the vorticity quantum given by:

$$\int Q \, dA = \hbar/mc \qquad (7.27)$$

where Q is the torsion (analogous to $\int B \, dA$ as we have the analogy between B and Q).

So now we have the uncertainty relation:

$$\Delta\Phi_m \cdot \Delta m \geq \hbar \qquad (7.28)$$

where $\Delta\Phi_m$ is in units of $\hbar/m$. In fact in liquid helium one observes the phenomenon of quantized vortices and, just as in a superconductor, one has quantized flux ($\hbar c/e$). As we have seen in section 4, we have analogy between Einstein-Cartan and London equation. In fact in our case we should also point out the analogy with the London equation which implies a modification of the electrodynamic equations inside a superconductor. For the London equation also we have a length parameter which gives the penetration depth which is the analogue of Q^{-1} here. For our case we have for instance, i.e. the torsion modified equation:

$$\nabla^2 H = (n\alpha/3\pi) H \cdot Q \qquad (7.29)$$

where the formal analogy with London's equation would again suggest Q^{-1} as the penetration depth, the photon again acquiring an effective mass, the important difference being that the London equation violates phase invariance wheras eq.(7.29) does not.

3. Spin, magnetism and vortices

Since torsion is also connected with magnetism [8], [9], we can invoke the critical magnetic field given by (see (4.60)):

$$B_{crit} = (m_\gamma^2 c^3/q\hbar) \qquad (7.30)$$

where q is an extra charge on the field as given by the additional term. This would also be consistent with the 'superfluid' background approach of Vigier [10].

Using for B_{crit} the primordial field of $\approx 10^{-7}$G and for $m_\gamma \approx 10^{-62}$g as estimated from the background torsion, we have a constraint on the extra charge as :

$$q \approx 10^{-48} e \qquad (7.31)$$

where e is the electric charge. This is much more stringent than the limits given by the isotropy of the cosmic ray background [11] which is $q \approx 10^{-32}e$. Thus this limit on the photon mass fixed by the background torsion, also constrains the photon charge to be $< 10^{-48}e$.

The well known dispersion relation for massive photons i.e. an effective refraction index $n \simeq (1 \quad (m_\gamma{}^2 c^4/2\hbar^2 f^2)$, would eg.for $m_\gamma \approx 10^{-62}$g imply a $(1 - v_\gamma^2/c^2) < 10^{-26}f^2$ and anisotropy changes of $\delta v_\gamma /c < 10^{-29}$, which are virtually unobservable.

Now the analogy between the viscous drag (Navier Stokes equation) [12] $\partial V/\partial t = \nu\nabla^2 V$, and the Schrödinger equation $\partial\psi/\partial t = (i\hbar/m)\nabla^2\psi$, (except for the factor i) suggests that for the quantized vortices we have the quantum viscosity given by $\hbar/m$. In fact $\hbar/m$ has the units of kinematic viscosity (cm^2s^{-1}). The lifetime of a vortex of scale L is given by $t \sim L^2/\nu$.

So if we picture particles with rest mass m as vortices (with the mass m arising from the quantized $\Delta\Phi_m = \int Q\ dA$ through eq.(7.28) just as charge follows from the flux (or magnetic moment) through eq.(7.26)) of scale length $L \simeq \hbar/mc$, then the quantized viscosity $\hbar/m$, (using the analogy of Schrödinger equation with drag equation) would imply a particle lifetime (since we have not inwoked the factor i, we have only damping, i.e decay and no oscillations) of:

$$t \approx L^2/v \simeq (\hbar/mc)^2/\hbar/m \simeq \hbar/mc^2 \qquad (7.32)$$

which is just the lifetime of the quantum particles. If hadrons are pictured as strong gravity black holes [13],[14],[15] then $L \approx G_f M/c^2$, and

$$t \approx (G_f m/c^2)^2/\hbar/m \approx G_f^2 m^3/hc^4 \qquad (7.33)$$

which is just the Hawking formula exactly for black hole evaporation time. For cosmic black holes, $G = G_N$ and the lifetime is $t = G^2M^3/\hbar c^4$; the lifetime is very long for astronomical black holes, because the corresponding quantum viscosity $\hbar/M$ is very small and the lifetime of the structure scales as L^2. For particles as m is very small, $\hbar/m$ is large and also the scale is small: so lifetime is very short. We have thus some kind of unified description through $\int Q\, dA$, the mass flux quantum. We may suggest some experiment with vortices in superfluid helium, a vortex of length L should last for $(L^2/\hbar/m)$ s, where m is the mass contained within the vortex. The time scale is bound to be very long since $\hbar$ is in the denominator. Also in neutron stars one should have very long lived superfluid vortices having both magnetic and mass fluxes.

There may be torsion effects in scattering of vortices with spins parallel and those with their spins antiparallel. Considering the analogy with scattering of polarized proton beams as discussed in [16], one would expect to see drastic differences in the two cases. This may help to resolve some of the anomalous phenomena associated with superfluid vortices.

The proton is stable because it is supported by torsion and strong gravity at the radius given by eq.(7.15) (see also [17]). As noted in [17], this is also some kind of universal relation also holding for macroscopic objects. Thus for a neutron star with 10^{57} nucleons, it gives $r \approx 10^6$cm = 10 km.

4. Spin in cosmology

We can also apply this equation in cosmology following the earlier work of Trautman [18]. The corresponding strong gravity cosmological framework is in refs.[6] [13]and [14]. This gives a value for the radius of the universe in the hadronic era, when the temperature is

$$T_h \approx m_p c^2/k_B \qquad (7.34)$$

The adiabatic expansion of the Universe as implied by the Robertson-Walker model would give:

$$RT = \text{constant} \qquad (7.35)$$

Now in the hadron era $T = T_h$ (eq.(7.34)); so $R_h T_h = R_0 T_0$ where R_0 and T_0 are present values. Eq. (7.15) for the radius of the proton can be written:

$$r_p \approx (3/8)^{1/3}(\hbar/m_p c) \qquad (7.36)$$

(since $G_f \approx \hbar c/m_p^2$). So the radius of the universe with $N \approx (\hbar c/G_N m_p^2)^2$ protons in the hadron era, can be written:

$$R_h = (3/8)^{1/3}(\hbar/m_p c)N^{1/3} \qquad (7.37)$$

(as follows from the exclusion principle). Each proton cannot have a radius smaller than r_p. So the universe behaves like a giant nucleus in the hadron era with a radius going like

$r_pA^{1/3}$, r_p= radius of individual proton, A = atomic mass ≈ number of nucleons):

$$R_h \approx (3/8)^{1/3}(\hbar/m_pc)(\hbar c/G_Nm_p^2)^{2/3} \qquad (7.38)$$

The present value of the radius of the universe is [18]:

$$R_0 \approx (\hbar c/G_Nm_p^2)(\hbar/m_pc) \qquad (7.39)$$

so from $R_hT_h = R_0T_0$:

$$T_h(3/8)^{1/3}(\hbar/m_pc)(\hbar c/G_Nm_p^2)^{2/3} = (\hbar c/G_Nm_p^2)(\hbar/m_pc)T_0 \qquad (7.40)$$

and using $T_h = m_pc^2/k_B$ (from eq.(7.34)) we have for T_0:

$$T_0 \approx (m_pc^2/k_B)(8/3)^{1/3}(G_Nm_p^2/\hbar c) = 2.73\ {}^0K \qquad (7.41)$$

which agrees very well with the recently measured value. In fact present data indicate that the cosmic background radiation is between 2.73 ^{0}K and 2.74 ^{0}K [19].

The factor 8/3 comes from torsion. This is important to have excellent agreement for T_0. At the Planck epoch since the particle masses are ≈ m_{Pl} , we have $T_{Pl}= m_{Pl}c^2/k_B$ and $Gm_{Pl}^2/\hbar c$ ≈ 1. At any epoch in the early universe, when energy ≈ mc^2 ≈ kT, the energy dependence of G (as implied by $G_fm_f^2 = G_wm_w^2 = G_Nm_{Pl}^2$) implies:

$$T \approx mc^2/k_B\ , \quad \text{when} \quad Gm^2/\hbar c \approx 1 \qquad (7.42)$$

So we seem to have a satisfactory description of the thermal history of the early universe. Now we will show in great detail that relating torsion to intrinsic unit of spin ħ, with the consideration that defects in space-time topology occur in multiples of Planck length, the electric charge appears as a derived concept showing its universality. Moreover, in this context, we cannot have a massless particle with charge (see Chapter VIII for quantum effects).

5. Charge and dislocations

Further it is interesting to note the analogy between the elastic deformation field around a dislocation line and the magnetic field of constant line currents. The current is replaced by the Burghers vector which must be constant along the dislocation lines like the current in electromagnetism.

Moreover the force between dislocations is a radial force with a potential of the form

$$V \sim \text{const.}\, b_1 b_2 / r \qquad (7.43)$$

(these are the couplings associated with the defects).

Dislocations of like sign (helicity) repel ($b_1 b_2 > 0$) while those of unlike sign (helicity) attract ($b_1 b_2 > 0$) [20]. This is strikingly similar to the behaviour of like and unlike charges in electromagnetism and support the possibility of interpreting charges as dislocations defects.

The differential expression for the law of conservation of the Burghers vector has the same form as the electric current:

$$\partial\rho/\partial t + (1/c)\partial j/\partial x = 0 \qquad (7.44)$$

Integrating over a surface spanning some closed line L we have:

$$db_k/dt = - \oint j_{ik} dx_i \qquad (7.45)$$

The integral on the right gives 'flux' of Burghers vector b_k through closed contour per unit time, i.e. the Burghers vector is carried across L by moving dislocations and j_{ik} may be called dislocation flux density.

Thus the dislocations of different helicity correspond to 'charges' and the moving dislocations give rise to a current, the line integral of the dislocation (charge) density giving the rate of change of this current with time.

6. Spin and fifth dimension

Now we like to discuss the Kaluza-Klein theory in the strong gravity framework: this leads naturally to the values of the elementary particle magnetic moments showing also that the spin of particle, through torsion, plays a more fundamental role then the electric charge that appears as a derived concept.

We have already seen the connection between strong

gravity and fundamental physical constants (see also [21],[22]).

In order to see how the Kaluza-Klein picture in the strong gravity framework also naturally leads to the values of the elementary particle magnetic moments, first of all we recall the old result that the Dirac wave equation usually written in a four-dimensional form can be written more naturally in a five-dimensional form, because of the existence of five anticommuting Dirac matrices. This five dimensional form when considered in the context of the Kaluza-Klein field theory, results [23,24] in a magnetic-moment term μ in four dimensions given by:

$$\mu = \hbar\sqrt{G}/2c \qquad (7.46)$$

where G is the gravitational constant. In the framework of strong gravity G is the strong gravity coupling constant G_f ($G_f \approx 6.6\cdot 10^{30}$cgs units $\approx 10^{38}$G) so that

$$\mu_f = \hbar\sqrt{G_f}/2c \qquad (7.47)$$

(related to the spin $\hbar/2$ of the particle). This gives for a typical hadron, a μ_f of $\sim 4\cdot 10^{-23}$erg/G, close to that of the anomalous magnetic moment of the proton ($\mu_p \approx 2.8\cdot e\hbar/2m_p c$).

As shown above, we had considered a finite classical theory of the electron with torsion and strong gravity, obtaining a finite radius of the electron as

$$r_{cl} = \left(\frac{3G_f \hbar^2}{8m_e c^4} \right)^{1/3} \sim 1.4 \cdot 10^{-13} \text{cm} \qquad (7.48)$$

(m_e is electron rest mass) when identified with the electron classical electromagnetic radius this gave a relation for the electric charge as:

$$e = (3G_f \hbar^2 m_e{}^2 c^2)^{1/6} \qquad (7.49)$$

This would enables us to write for the Bohr magneton the expression:

$$\mu_B = \frac{(3G_f)^{1/6}}{2} \left(\frac{\hbar^4}{m_e{}^2 c^2} \right)^{1/3} \qquad (7.50)$$

It would be noticed that the electron charge e does not enter in the formulae (7.47),(7.48), and (7.50). In fact in ref.[25], the electric charge is a derived concept, with the spin $\hbar/2$ of the particle playing a more fundamental role entering through the torsion term and providing a finite classical radius by balancing the strong gravitational force between the constituents. Indeed the equation (7.47) is basic in the sense that replacing m_e with the proton mass m_p, we get a finite classical radius for the proton as $r_p \approx 1.2 \cdot 10^{-14}$cm. Here again the balance is between strong gravity and torsion i.e.

$$r_p = (3G_f \hbar^2/8m_p c^4)^{1/3} \sim 1.2\cdot 10^{-14}\text{cm} \qquad (7.51)$$

If we identify r_p now with the strong interaction radius $\approx g^2/2m_p c^2$ (g^2 is the strong coupling charge squared), then we have the relation:

$$g^2/e^2 = (m_p/m_e)^{2/3} \qquad (7.52)$$

which with $e^2/\hbar c \approx 1/137$ would give for the strong coupling $g^2/\hbar c \approx 1$, i.e. the strength of the strong interaction. Thus with just torsion entering through the spin of the particle, we are able to obtain both electromegnetic and strong coupling constants through G_f and the particle mass m_p , m_e. We can also connect all this with what was said in Chapter VI wherein the need for an energy dependent torsion coupling constant (going as E^{-2} with energy) in the early universe was emphasized, i.e. we had the relation (see also eq.(7.24) of the present Chapter)

$$G_f\, m_p^{\;2} = G_N\, m_{Pl}^{\;2} = G_w\, m_w^{\;2} \qquad (7.53)$$

(m_{Pl}, m_w, are the Planck and weak intermediate boson masses, $G_w = G_F(c/\hbar)^{1/2}$, with G_F and G_N the Fermi and Newtonian constants). Thus for masses (energies) of the Planck mass, the coupling becomes G_N, the Newtonian value. Then eq.(7.51) gives the radius

$$\left(\frac{3G_N\hbar c^2}{8m_{Pl}c^4}\right)^{1/3} \approx 1.5\cdot 10^{-33}\text{cm} \qquad (7.54)$$

which is the Planck length for particle of Planck mass, and for the intermediate boson mass, we get the beta decay length $\sim 6\cdot 10^{-17}$cm giving the corresponding radius for the boson.

So in the early universe the magnetic moments and other attribute of the elementary particles are ''frozen'' inside them. As neutrinos take part only in weak interactions with a strength $\sim (G_F/\hbar c)\left(\frac{m_e c^2}{\hbar}\right)^2 \sim 10^{-12}G_F$, we would expect a neutrino magnetic moment (from eq.(7.46)) of $\mu_\nu \approx 10^{-9}\mu_B$, to be ''frozen in'' during the lepton era. A neutrino magnetic moment of this order has indeed been invoked to explain the solar neutrino puzzle. We stress that here we obtain this value of μ_ν for <u>massless</u> neutrinos (the spin $\hbar/2$ of the neutrino being the sole attribute leading to μ in eq.(7.47)). Alternatively we would expect the weak interaction to give corrections to the electron and muon magnetic moments of magnitude $\approx 10^{-9}\mu_B$. This could be a manifestation of the Klein-Kaluza effect in five dimensions and could be tested (currently the magnetic moments are not known to that accuracy!) The limit of $\mu_\nu \approx 10^{-9}\mu_B$ is barely consistent with electron-neutrino scattering data. In this connection see also possible temperature dependent effects in early universe [26].

If the particle spin is the more basic attribute, we can probably visualize spin as the fifth extra dimension in the Klein-Kaluza theory rather than the electric charge. The theory can be modified to include this. We give a brief outline.

In the context of a gauge theory of torsion we can write the Einstein-Cartan Lagrangian as:

$$L = \int d^4x\sqrt{-g}\left(-\frac{R}{2\chi} - \frac{1}{4\chi} S^{\mu\nu\beta}S_{\mu\nu\beta}\right) \tag{7.55}$$

where the antisymmetric field strength $S_{\mu\nu\beta}$ can be written in terms of the gauge field strength as [6]:

$$S_{\mu\nu\beta} = c_\mu{}^{\times} \; f_{\nu\beta} \; ; \qquad f_{\nu\beta}\cdot\bar{g} = F_{\nu\beta} \quad \text{etc.} \tag{7.56}$$

where f and F are field strengths and $\bar{g}$ is the determinant of the group metric.

Analogous to electromagnetism, $S_{\mu\nu\beta}$ satisfies the identity:

$$\partial_\rho\left(\varepsilon^{\rho\mu\nu\sigma} S_{\mu\nu\sigma}\right) = 0$$

Moreover through a duality transformation: $S_{\mu\nu\beta} = \varepsilon_{\mu\nu\sigma\rho}\partial^\sigma s$, the Lagrangian just becomes:

$$L = \int d^4x \sqrt{-g}\left\{-\frac{R}{2\chi} + (1/2)\partial^\mu s \; \partial_\mu s\right\} \tag{7.57}$$

where s is the spin scalar. This action in four dimensions involving curvature and spin can be written as just curvature in 5 dimensions, i.e.:

$$R^5 = R^4 + (1/2)\partial^{\mu}s\delta_{\mu}s \qquad (7.58)$$

Then the fifth dimension would be the spin of the particle rather than electric charge. For quantization of the spin, we note that supposing the probability density not to depend on the fifth dimension (any co-ordinate transformation on the fifth is just a translation) then the form of the state vector for an interference experiment with two or more paths for a particle has the form:

$$\psi = \exp[ikx_5']\psi' + \exp[ikx_5]\psi \qquad (7.59)$$

where k is some observable. To have invariance of the interference pattern, $k(x_5' - x_5) = 2\pi n$. Thus the periodicity of the fifth co-ordinate $x_5' - x_5 = 1/2\pi$ would ensure the quantization of the observable k which could be the spin quantized in units of $\hbar$ rather than the electric charge. In this case, the radius of the compact space would be just the Planck length. In fact usually for Klein-Kaluza unification, the radius of the compact space (the unit circle for electromagnetism) is given by $R_c = \frac{R_{Pl}}{\alpha_g^{1/2}}$, where α_g for electromagnetism is just the fine structure constant. In general $\alpha_g \simeq g^2/\hbar c \ll 1$, so $R_c > R_{Pl}$; but with spin replacing the charge, $g^2 \approx \hbar c$, so $R_c \approx$ Planck length. Another support for the fundamentality of spin comes from tha fact that eq. (7.45) for the magnetic moment seems to be a universal relation for all macroscopic bodies as well, in the form $\mu = \beta(J/c)\sqrt{G}$, (J is the intrinsic angular momentum and β a constant ≈ 1) (see Chapter II and ref.[27]).

References

[1] H.Lorentz - Theory of electron, Dover, N.Y.1952
[2] M.Born - Proc.Roy. Soc. A143, 410 (1934)
[3] M.Born and L.Infeld - Proc.Roy.Soc. A144, 425 (1934)
[4] H.Casimir - Physica 19, 848 (1956)
[5] V.de Sabbata and M.Gasperini - Nuovo Cimento Lett. 24, 520 (1979)
[6] C.Sivaram and K.Sinha - Physics Report 51, 111 (1979)
[7] V.de Sabbata and C.Sivaram, Astrophys.Space Sci. 158, 347 (1990)
[8] V.de Sabbata and M.Gasperini - Lett.Nuovo Cimento 27, 133 (1980)
[9] V.de Sabbata and C.Sivaram - Il Nuovo Cimento 102B, 107 (1988)
[10] J.P.Vigier et.al. - Phys. Letters 154A, 203 (1991)
[11] C.Sivaram - Progress Theor.Phys.82, 115 (1989)
[12] see for instance Landau et Lifshitz - Mécanique des fluides editions de Moscou (1971) pg.65
[13] V.de Sabbata and P.Rizzati - Lett.Nuovo Cimento 20, 525 (1977)
[14] V.de Sabbata and M.Gasperini - Lett.Nuovo Cimento 25, 489 (1979)

[15] V.de Sabbata and C.Sivaram - 'Torsion and Quantum effects' in vol. devoted to 85th jubilee of D.D.Ivanenko, "Modern Problems of Theoretical Physics" ed.by P.I.Pronin and Yu.N. Obukhov, World Sci.Singapore 1991 pg.143

[16] V.de Sabbata and C.Sivaram - Il Nuovo Cimento 101A, 273 (1989)

[17] V.de Sabbata and C.Sivaram - Il Nuovo Cimento 105B, 1181 (1990)

[18] A.Trautman - Nature 242, 7 (1973)

[19] R.A.Alpher and R.Herman Physics Today 41, 8 (1988) and also 'Particle Properties data Booklet' pp.163-164 (1988)

[20] Landau and Lifshitz - Theory of Elasticity, section 27, 29 pg.34 (Pergamon Press, 1962)

[21] V.de Sabbata and C.Sivaram - Ann.der Physik 47, 511 (1990)

[22] V.de Sabbata - in "Gravitational Measurements, Fundamental Metrology and Constants" ed.by V.de Sabbata and V.N.Melnikov, Kluwer Acad.Publ.1988, pg.115

[23] W.Pauli - Ann.der Phys. 18, 337 (1933)

[24] A.O.Barut and T.Gormitz - Found.Phys. 15, 433 (1985)

[24] V.de Sabbata and C.Sivaram - Found.Physics Letters 6, 201 (1993)

[25] V.de Sabbata and C.Sivaram - Nuovo Cimento 105B, 603 (1990)

[27] V.de Sabbata and C.Sivaram - Nuovo Cimento 100A, 919 (1988)

CHAPTER VIII

TORSION: A WAY TOWARD QUANTIZATION OF GRAVITY?

1. Universality of electric charge

In the previous chapter we have seen that the electric charge e was sought to be understood as a derived concept with the spin $\hbar/2$ of the particle playing a more fundamental role entering through the torsion term and providing a finite classical radius by balancing the strong gravitational force. We had also obtained expressions for the Bohr magneton and electron classical radius which did not involve the electric charge [1], [2], [3]. In fact from equation (7.14) we can write the expression for the Bohr magneton as

$$\mu_B = (3G_f/2)^{1/6}(\hbar^4/m_e^2c^2)^{1/3} \qquad (8.1)$$

We have also extended the concept to particles of different masses which may also undergo other interactions (like strong interactions in the case of hadrons, or weak interactions in the case of W-bosons). It is an established empirical fact that all elementary particles irrespective of their masses or interactions undergone, carry the same electric charge; i.e. electrons, protons, W-bosons etc. all have the same value of electric charge e.

Here we shall attempt to account for this basic fact by interpreting charge as the total flux (in the sense of Gauss's theorem) associated with the space-time defect induced by torsion. If one treats space-time as a deformable medium with elastic properties one can use the analogy of the geometrical description of crystal dislocations or defects where it is known that torsion plays the role of defect density in the limit of dislocations having a continuous distribution. We consider a small closed circuit and write

$$l^{\alpha} = \oint Q^{\alpha}_{\beta\gamma} dA^{\beta\gamma} , \qquad (8.2)$$

where $dA^{\beta\gamma} = dx^{\beta} \wedge dx^{\gamma}$ is the area element enclosed by the loop and $Q^{\alpha}_{\beta\gamma} = \Gamma^{\alpha}_{[\beta\gamma]}$ is, as usual, the torsion associated with the connection. Here torsion has an intrinsic geometric meaning, it represents the failure of the loop to close (analogously to the crystal case), l^{α} having the dimension of length. Torsion can be related to intrinsic unit of spin $\hbar$, by postulating that the defects occur in space-time topology in multiples of the Planck length, $(\hbar G/c^{3})^{1/2}$.

We have shown elsewhere (see equation (2.27)) that the spin-torsion coupling induces a net magnetic moment and a magnetic field given by:

$$B = (8\pi/3c)(\alpha G)^{1/2}\sigma \qquad (8.3)$$

where σ is the spin density associated with the defect, i.e. σ is now given by:

$$\sigma = \hbar \Big/ \left\{ 2(4\pi/3)(\hbar G/c^3)^{3/2} \right\} \qquad (8.4)$$

$(4\pi/3)(\hbar G/c^3)^{3/2}$ being the volume associated with the defect. Substituting in eq.(8.3) we have

$$B = \left(c^{7/2} \Big/ G\hbar^{1/2} \right) \alpha^{1/2} \qquad (8.5)$$

So the total flux associated with the space-time topological defect induced by torsion is given by

$$\int B \cdot dA = \left(c^{7/2} \Big/ G\hbar^{1/2} \right) 4\pi(\hbar G/c^3)\alpha^{1/2} = 4\pi e \qquad (8.6)$$

(which by Gauss's theorem implies a charge e associated with the flux).

The magnetic charge and the electric charge are related by a numerical factor through the Dirac quantization condition which again implies universality of the electric charge. The magnetic induction is related to the spin coupling through torsion of space -time. This provide a finite radius for the electron, where torsion supported the electron against collapse providing a finite energy and a finite radius equal to the elctron classical radius. The radius was obtained as (see equation (7.13)):

$$r_{cl} = (3G_f \hbar^2/8m_e c^4)^{1/3} \approx 1.4 \cdot 10^{-13}\text{cm} \qquad (8.7)$$

which is an exact solution of the spin modified Poisson equation:

$$\nabla^2\Phi = 4\pi G\,(\rho - G\sigma^2) \qquad (8.8)$$

where σ is the spin density; so this provides the appropriate distance. The electric charge is obtained, as we have seen in equation (7.14) by:

$$e = (3G_f\hbar^2 m_e^2 c^2)^{1/6} \cong 5\cdot 10^{-10}\text{esu} \qquad (8.9)$$

Notice that the constant α appears in eq.(8.3) (and then in eq.(8.6)) only from consideration of magnetic moment obtained from torsion and also from consideration of spin as the fifth dimension (rather than electric charge) in the Kaluza-Klein [2] framework. The Lagrangian then becomes that of the Einstein-Cartan theory, implying a magnetic moment from just the spin of particle i.e.

$$\mu = \hbar\sqrt{G}\,/\,2c \qquad (8.10)$$

In case of the electron this was shown to give the Bohr magneton. In all these cases, the charge enters as a derived concept, with the spin of the particle playing a more fundamental role. In the case of Dirac's theory, charge quantization occurs through the relation $eg = \hbar c$, 'e' and 'g' being the electric and magnetic charges. Thus a system consisting of an electric and magnetic charge automatically acquires an angular momentum or intrinsic spin $\hbar$. Here instead we start with the spin $\hbar$ as fundamental which, when built into the structure of space-time gives rise to torsion and a magnetic moment leading to the electric charge. So once again we have universality of charge and $\hbar$. Instead of Dirac quantization condition, we have the corresponding gravitational analogue given by eqs.(8.11) and (8.13). We had obtained the classical electromagnetic radius and magnetic moment from the gravitational charge. The quantization of the gravitational charge (i.e. the direct analogue of the Dirac relation in general relativity) was obtained also by Motz. So here we have the electric charge arising from gravitational charge which in turn arises from defects in space-time topology induced by torsion caused by the existence of the fundamental unit of spin '$\hbar$'. So we start with $\hbar$ and obtain the gravitational and electric charge; we have no need for a monopole in our framework.

It would be noticed that the above relation for flux (eq.(8.6)) is independent of G. In fact in order to describe the different interactions, we had invoked an energy dependent G, G scaling with energy as E^{-2} (see chapter VI). The defect length l^{α} would then scale as E^{-1} . At 100 Gev it would be the beta-decay length, at 10^{19} Gev the Planck length, at 1 Gev the proton Compton length and so on. These would correspond to particles of masses $\approx$ 250 Gev (mediating the weak interactions at high energy i.e. w'-bosons), ≈ 1 Gev (the proton and f-mesons), Planck mass at 10^{19} Gev etc. So that the energy dependent torsion coupling gives an energy dependent defect length l^{α} (corresponding to the scales at which different interactions operate). However whatever be the defect length or interaction scale or particle mass, the intrinsic spin would always be ħ as:

$$G_N m_{Pl}^2 \approx G_w m_w^2 = G_f m_f^2 = \hbar c \qquad (8.11)$$

(the Fermi beta-decay constant G_F is related to the effective gravitational constant describing weak interactions i.e. G_w through $G_w = G_F(c/\hbar)^2$).

Also at the GUTs energy scale, $G = G_{GUT}$ we have $G_{GUT}m_{GUT}^2 = \hbar c$. Eq.(8.6) now shows that whatever be the defect size or interaction scale, the electric charge now interpreted as the flux associated with the defect is always e. So e becomes another fundamental constant with ħ and c, which is same for particles of all masses.

So whereas previously our eq.(8.11) gave a natural basis for gravitational, weak and strong interactions (interpreted through an effective energy dependent G as described by eq.(8.11)), we now also have an explanation for the universality of the electric charge e (i.e.electromagnetism) which is independent of G and is the same for all particles like their intrinsic spin. So particles in this picture correspond to defects induced in space-time by energy dependent torsion coupling constant. Particles like neutrinos or photons which are massless carry no charge and are <u>not localized</u> in space-time. Defects have to be localized in space-time. So since neutrinos or photons are not defects, they do not carry electric charge (one cannot define a Gauss's theorem over a closed surface characterizing the defect, so that there is no flux or charge).

So definitely we cannot have a massless particle with charge. Also no stable fermion having mass but no charge is known. For unstable particles like neutrons or massive bosons, the situation can be expected to be more complicated.

The universality of electric charge is also apparent (but not explicity stated) in another work [4] where an alternative to the Higgs mechanism was proposed to generate particle masses. In particular the analogy with superconductivity was used to generate an effective mass for the electroweak-boson below the critical electroweak magnetic field induced by torsion. The intermediate boson mass obtained was [4], [5]:

$$m^2c^3/e\hbar = (8\pi/3c)(2\alpha G\ )^{1/2}\left\{\hbar/(G\ /\hbar c)^{3/2}(4\pi/3)\right\}(c/\hbar) \quad (8.12)$$

and using $G_F = G_w(\hbar/c)^2$ this gives $m_z = e/\sqrt{G_w}$, which gave $m_z \approx 70$ Gev, agreeing well with the observed mass of electroweak boson.

When the appropriate strong gravity constant G_f is used in eq.(8.12) we get the muon mass as $m_\mu \approx 105$ Mev; for the Newtonian constant G_N, eq.(8.12) would give the mass as $e/\sqrt{G_N}$. Equation (8.12) would once again suggest a universal scaling relation:

$$G_N m_\alpha^2 = G_w m_z^2 = G_f m_\mu^2 = e^2 \qquad (8.13)$$

which differs from the relation (8.11) by a factor $\alpha^{1/2}$ where α is the fine structure constant.

Such a relation also arises naturally whenever unification of electromagnetism with gravitation or weak or strong interaction is considered. Thus unification with newtonian gravity gives $G_N m_\alpha^2 = e^2$, with weak interaction $G_w m_z^2 = e^2$, and with strong interaction (through the strong gravitational constantG_f) $G_f m_\mu^2 = e^2$.

In addition eq.(8.12) also implies that the electric charge e is always the same for all these different energy scales, only the effective G scales with energy; e and the intrinsic spin ħ as suggested by eqs.(8.11) and (8.13) respectively are always the same at all energy scales and for all particles.

2. Torsion and defects in space-time

Now we will like to deepened the question of defects in space-time topology that until now we have used to show the analogy between torsion and magnetic field (see eq.(7.26),(7.27)) and also the analogy with London equation (see (7.29)), and will explore the possibility to go toward the quantization of the space-time itself.

In fact we will show that introducing torsion in general relativity, that is, physically, considering the effect of the spin, and linking the torsion to defects in space-time topology we can have a minimal unit of time. Also an uncertainty relation between time and temperature is suggested. The interesting thing is that with this minimal time we can eliminate the divergence of self-energy integral without introducing any ad hoc cut-off and it is also possible to understand black hole evaporation as a proceess of quantum diffusion which leads directly to the Hawking formula. A minimal operationally definable temperature in a cosmological context is discussed.

One of the major difficulties in contemporary physics both in quantum field theory and in general relativity is the presence of singularities or infinities. In general this depends on the fact that we are used to considering point particles so that we are faced with the divergence of self energy integrals that go to infinity, or when in general relativity we have to do with a collapsing body (i.e. a black hole) we are faced with the singularity in $r \longrightarrow 0$.

These difficulties may disappear if we have not only a minimal length, but also a minimal unit of time, that is if we have to do with a discretized space-time. In fact, for instance, we know that if we introduce torsion in general relativity (that means the consideration not only of mass but also of the spin), we can avoid singularity when considering the collapse of the Universe (Trautman). That is we have a minimum radius of the universe. The problem is then to have a similar behaviour not only for spatial dimension but also for time.

For that we will consider in detail the connection of torsion with defects in space-time topology.

We know that in the geometrical description of crystal dislocations and defects, torsion plays the role of defect density (in this context we consider space-time as an elastic deformable medium in the sense of Sakharov).

If we consider a small closed circuit and write:

$$l^{\alpha} = \oint Q^{\alpha}_{\beta\gamma} dA^{\beta\gamma} , \qquad (8.14)$$

where $dA^{\beta\gamma} = dx^{\beta} \wedge dx^{\gamma}$ is the area element enclosed by the loop and $Q^{\alpha}_{\beta\gamma} = \Gamma^{\alpha}_{[\beta\gamma]}$ is as usual the torsion associated with the connection $\Gamma^{\alpha}_{\beta\gamma}$, then l^{α} represents the closure failure, i.e. torsion has the intrinsic geometric meaning of the failure of the loop to close, analogous to the crystal case, l^{α} having the dimensions of length. In the above equation torsion can be related to the fundamental unit of intrinsic spin $\hbar$, by postulating that defects in space-time topology at the quantum level should occur in multiples of the Planck length $(\hbar G/c^3)^{1/2}$; i.e. we have

$$\oint QdA \cong n(\hbar G/c^3)^{1/2} \quad (8.15)$$

Time would thus be defined in the quantum geometric level through torsion as

$$t = (1/c)\oint Q\, dA = n(\hbar G/c^5)^{1/2} \quad (8.16)$$

so torsion is essential to have a minimum unit of time $\neq 0$! This in fact would give us the smallest definible unit of time as $(\hbar G/c^5)^{1/2} \cong 10^{-43}$s. In the limit of $\hbar \Longrightarrow 0$ (classical geometry of general relativity) or $c \Longrightarrow \infty$ (Newtonian case), we would recover the unphysical $t \Longrightarrow 0$ of classical cosmology or physics. So both $\hbar$ and c must be finite to give a geometric unit for time (i.e. $\hbar \Longrightarrow 0$ and $c \Longrightarrow \infty$ are equivalent). The fact that $\hbar$ is related to a quantized timelike vector, discretize time. This quantum of time or minimal unit of time also correspondingly implies a limiting frequency of $f_{max} \approx$ $\approx (c^5/\hbar G)^{1/2}$. This would have consequences even for perturbative QED, in estimating self energies of electrons and other particles, i.e. the self energy integral (in momentum space) taken over the momenta of all virtual photons. To make the integral converge, Feynman in his paper on QED [6], multiplied the photon propagator, k^{-2}, by the ad hoc factor: $-f^2/(k^2 - f^2)$, where k is the frequency (momentum) of the virtual photon. This convergence factor although it preserves relativistic invariance, is objectionable because of its ad hoc character without any theoretical justification. Feynman considers f to be arbitrarily large without definite theoretical basis. Here the presence of space-time defects associated with the torsion due to the intrinsic spin would give a natural basis for the maximal value for f^2_{max} as (from eq.(8.16)) $\approx c^5/Gh \approx 10^{96}$ (and extremely large as required by Feynman), giving finite result (instead of ∞) for the self energy. This makes f_{max} another fundamental constant for particle physics serving as a high frequency cut off which is not arbitrary.

In the previous pages we had pictured particles with rest mass m as vortices, which would give them a life time related through torsion as (i.e. analogous to charge being connected to $\int B\, dA$ we had mass related to $\int Q\, dA$ exploiting analogy between torsion and magnetism):

$$t \cong (1/c)\int Q\, dA \cong h/mc^2 \qquad (8.17)$$

which is just the lifetime of quantum particles.

We can understand the connection between eqs.(8.16) and (8.17) by having an energy-dependent G! as we have seen in chapter 6. We have in fact:

$$G \approx \hbar c/m^2 \qquad (8.18)$$

For $m = M_{Pl}$, we have $G = G_N$. For strong gravity with $m = m_p$, we have $G = G_f$ and so on. Substituting (8.18) in (8.16), we arrive at eq.(8.17), and we have space-time defect lengths l^α scaling with energy as E^{-1}, where the strength of interaction is fixed by the value of G_{eff} related to l^α through eqs.(8.14), (8.16), (8.17) and (8.18). G_{eff} can be defined as $G_{eff} = G_N(M_{Pl}/M)^2$ {see section 1 of Chapter VI and also section 2 of Chapter XI (eq.(11.30))}.

The main point that must be stressed is that the existence of the different defect lengths l^α is primary and is due to the introduction of torsion that physically means the introduction of spin, and then becomes related to $\hbar$, the unit of spin.

So the absence of torsion implies absence of defects in space-time and consenquently absence of masses, charges and of the very existence of time!

3. Temperature, time and torsion

Now $\hbar$ is related to time through the energy, that is we have:

$$\text{time} \Longrightarrow \hbar/\text{energy} \quad \text{i.e. we have } \hbar\nu = \text{energy} = E \qquad (8.19)$$

We have also the thermodynamic definition

$$E = k_B T \qquad (8.20)$$

where k_B is the Boltzmann constant (the unit of entropy: see below) and t is the temperature. Eqs.(8.19) and (8.20) imply the relation:

$$\text{time} \times \text{temperature} = \hbar/k_B = \text{constant} \qquad (8.21)$$

i.e. we have

$$\Delta t\, \Delta T = \hbar/k_B = 10^{-27}/1.3\cdot 10^{-16} \approx 10^{-11} \qquad (8.22)$$

Eq.(8.22) is universally valid as can be seen by several examples. In the early universe, for $\Delta t = 10^{-43}$s, we have $\Delta T \approx 10^{32}$K (from eq.(8.22)). This is just the temperature $T \approx (\hbar c^5/G_N)^{1/2}(1/k_B) \approx 10^{32}$K, at the Planck epoch. At the hadron era, $\Delta t \approx 10^{-23}$s, $\Delta T \approx 10^{12}$K, the hadronic temperature $(m_p c^2/k_B \approx 10^{12}$K) etc. $\approx (\hbar c^5/G_f)^{1/2}(1/k_B)$, Also for a typical electromagnetic interaction time scale $\approx 10^{-16}$s, we have the corresponding temperature from eq.(8.22) as $T = 10^5$ K, which is just that corresponding to the Rydberg theory ($\approx$ 13 ev) for ionized atoms.

So we can say that the existence of these times and temperatures in the early universe (as implied by eq.(8.22)) fixes the strength of the dominating interactions at the different epochs, G_N at t_{Pl}, G_f at t_{hadron} etc. i.e. G is automatically determined, once eq.(8.22) is assumed (through eqs.(8.16), (8.17) and (8.18)).

The time-temperature uncertainty relation eq.(8.22) is valid also in curved space-time. In general in curved space-time, the temperature T is modified as:

$$T(g_{oo})^{1/2} = \text{const.} \qquad (8.23)$$

But we also know that frequencies are redshifted in a gravitational field i.e.

$$\nu(g_{oo})^{1/2} = \text{const.} \qquad (8.24)$$

This is dilated in a gravitational field as $t(g_{oo})^{-1/2}$ (in general $\Delta\nu\Delta t = 1$!). So we have:

$$\Delta T(g_{oo})^{1/2}\cdot\Delta t(g_{oo})^{-1/2} = \Delta T\Delta t = \hbar/k_B = \text{constant} \qquad (8.25)$$

So the time-temperature uncertainty relation also holds in curved space and $\hbar$ and k_B are universal constants even in the presence of gravitation.

This happens to be a new uncertainty principle, the time-temperature uncertainty relation, which holds also in curved space-time. The minimal unit of time appears to be very promising because, as we have seen, we can eliminate the infinite divergence of self energy integral that until now was resolved only with some ad hoc cut-off.

4. Consequences for black hole physics

Moreover another important result is found when considering the collapse and evaporation of a black hole. In fact evaporation and Hawking formula comes out quite naturally owing to the existence of a minimal unit of time (t_{Pl} in this case).

In fact for an observer falling into the black hole towards the centre there is no longer the singular behaviour of classical general relativity. The minimal length and time are l_{Pl} and t_{Pl} (and not zero). However for an outside observer all events within the horizon are inaccessible. So his smallest interval is $t_{hor} \approx 2GM/c^3 \approx 10^{-5}$s for a solar mass black hole. The redshift factor is no longer ∞ but $GM_{\odot}/c^3(\hbar G/c^5)^{1/2} \approx 10^{38}$ owing to the existence of a minimal time.

As far as the behaviour of quanta within the horizon is concerned, we can consider them to diffuse inside the horizon with a mean free path $\approx l_{Pl}$ (i.e. matter does not collapse to ∞ density). Similar to the case of radiation diffusion within an ordinary star, we can consider the radiation inside the event horizon to random walk. So by the analogy to the case of an ordinary star, the diffusion length is estimated as follows: the total number of scatterings is given by:

$$\sqrt{N} = R_S/l_{Pl} \approx (GM/c^2)/(\hbar G/c^3)^{1/2} \qquad (8.26)$$

$$N \approx (GM/c^2)^2 \cdot c^3/\hbar G = (G^2M^2/c^4)(c^3/\hbar G) = GM^2/\hbar c$$

Total distance travelled by a quantum of radiation is

$$N \cdot l_{Pl} = (GM^2/\hbar c)\cdot(\hbar G/c^3)^{1/2} \qquad (8.27)$$

As measured by an observer inside the horizon, the time taken for the radiation to travel a distance R_S is given by

$$t_S = (GM^2/\hbar c)(\hbar G/c^5)^{1/2} \qquad (8.28)$$

(note the quantum of time enters the formula).

For an observer outside the horizon, this would be multiplied by the redshift factor found earlier, i.e. $GM_\odot/c^3(\hbar G/c^5)^{1/2}$, so that we have the time scale for the radiation to leak out of the black hole as

$$t_h = G^2M^3/\hbar c^4 \qquad (8.29)$$

which is the same as Hawking's formula!

As regards the minimal temperature it can be related to minimal possible energy allowed in the cosmological context. Also from time-temperature uncertainty relation $\Delta t \Delta T \approx \hbar/k_B$ we have that the maximal conceivable time is related to the Hubble H_o constant as:

$$\Delta t_{max} \approx 1/H_o \tag{8.30}$$

and then

$$\Delta T_{min} \approx \hbar H_o/k_B \approx 10^{-27} \cdot 10^{-18}/10^{-16} \tag{8.31}$$

So the smallest possible operationally definable temperature is

$$\Delta T_{min} \approx 10^{-29} \text{ K} \tag{8.32}$$

Here $\hbar H_o$ can have the interpretation as the minimum amount of energy that can be operationally defined in a closed universe [6].

In general one cannot reach absolute zero. One can come arbitrarily close (at present we have reached $\approx 10^{-6}$ K). It is remarkable that this minimal temperature can also be arrived at by considering a black hole of maximum possible mass, i.e. mass of the universe $\approx 10^{55}$g $\approx 10^{60} M_{Pl}$. Since temperature of a black hole scales inversely as mass: $T \approx \hbar c^3/GK_B M$, maximal possible mass M_{max} will give minimum possible operationally definable temperature which is consistent with (8.32)! Also from entropy considerations we have the maximal possible entropy of $\approx 10^{120} k_B$ [7]. This implies minimum temperature of $\approx 10^{-29}$K.

5. The entropy problem for evaporating black holes

Now we like to consider in more detail the question of minimum entropy.

Are well known the considerations of Hawking [8] about the evolution of quantum field in background metric of classical black hole. He finds that particle creation takes place and his semiclassical treatment of evaporation is valid if Schwarzschild radius >> Compton wave length i.e. $GM/c^2 \gg \hbar/Mc$, or $GM^2/(\hbar c) \gg 1$.

This quantity is also proportional to the entropy or area of the black hole. Emitted particles build a thermal spectrum and are uncorrelated among themselves. As black hole loses mass, its temperature rises, and evaporation accelerates (the rate is proportional to M^{-2}) as mass approaches the Planck mass $= (\hbar c^5/G)^{1/2} \approx 10^{19}$Gev). However at this stage the Compton length is comparable to the Schwarzschild radius, i.e. $GM/c^2 \approx \hbar/Mc$ for $M = M_{Pl}$, i.e. $GM_{Pl}^2/\hbar c \approx 1$, so that a quantum treatment of the gravitational field is required. The semiclassical approach would suggest explosive decay of black hole of mass M_{Pl} (temperature $\approx (1/k_B)(\hbar c^5/G)^{1/2} \approx 10^{32}$) in a burst of duration $(\hbar G/c^5)^{1/2} \approx 10^{-43}$sec. In the absence of any conservation law prohibiting the decay one might expect decay into various elementary field quanta.

This explosive disappearance when M reaches M_{Pl} would leave behind residue of thermal radiation which is a mixed state, in quantum mechanical term, i.e. the initial pure state has evolved into a mixed state of thermal radiation as seen by an observer at infinity. Such conversion would violate the fundamental tenets of quantum mechanics in flat space such as loss of quantum coherence (i.e. losses of phase between in and out states); it appears also inconsistent with quantum unitary postulate: "time evolution is governed by unitary operator in Hilbert space of states" which will be maintained in a pratically stable remnant. The emitted quanta are an uncorrelated ensemble and cannot be described by a single wave function; some information about emitted radiation is contained inside event horizon in the form of correlations between photons absorbed by black holes and those emitted to infinity; i.e. unitary postulate is maintained by correlating each state ψ_n of radiation with a corresponding state Φ_n of black hole so that joint system has a well defined overall wave function $|\psi_{tot}\rangle = \sum A_n |\psi_n^{rad}\rangle \times |\phi_n^{bh}\rangle$ that is we have a quantum state inside a black hole combined with thermal state of radiation to construct a pure state $|\psi_{tot}\rangle$ so that the unitarity is maintained.

Hawking process consists in pair creation in a strong gravitational field, that is one member is emitted to infinity and the other is staying near horizon. When the degrees of freedom of the black hole are integrated out, external radiation is described by a density matrix $\bar{\rho}_{rad} = \sum|A_n|^2|\psi_n^{rad}\rangle\langle\psi_n^{rad}|$ (only stationary states found for a black hole are consistent with the statistical nature of thermal radiation suggesting continous random emission of thermal radiation). If the residual mini black hole disappears completely by decaying into ordinary particles we will be deprived of the large reservoir of states in the black hole which is required for the construction of the overall wave function; the whole system will be described by the density matrix and the unitarity postulate is violated. Again as the initial and final states do not contain black holes we have to abandon even the weak assumption that an S-matrix exists between arbitrary initial and final asymptotic states that is as the black hole evaporates, the entropy $\propto (GM^2/\hbar c)\cdot k_B$, decreases continously. In order to preserve unitarity and coherence, residual mini black hole should not decay completely. Stable remnant should survive. Final state is not known but it is reasonable to assume from time reversal invariance that final state consists of particles not more exotic than the ones that made the black hole in the beginning.

There are essentially three possibilities:

a) the final state of evaporation may leave behind a naked singularity. This entails violation of cosmic censorship at quantum level [8],[9]

b) black hole may evaporate completely leaving no residue giving rise to serious problem with quantum consistency (coherency) described above; moreover by CPT theorem if there is no singularity initially, system must return to state with no singularity

c) stable remnant or residue of around Planck mass might remain; semiclassical back reaction and surface correction terms suggest that emission process might stop, but still better, and surely more physical, one may consider the possibility of a M_{Pl} black hole with spin ($\hbar$).

We think and we will see that just this third possibility is perhaps the more probable.

For a black hole with $M = a$, $a = s/Mc$, s being total spin, the black hole temperature is given by [10]

$$T_{bh} = \frac{\hbar^2 (M^2 - a^2)^{1/2}}{32 \pi K_B M [M + (M^2 - a^2)^{1/2}]} \tag{8.33}$$

and can be zero.

For $M = M_{Pl}$, indeed, $GM_{Pl}/c^2 \simeq \hbar/M_{Pl}c$, i.e. $M = a$

that is $T_{bh} = 0$ for Planck mass black holes provided they have a spin of $\sim \hbar$. So in that case black hole may indeed be stable! With the attainment of zero temperature such black holes would stop radiating and may be stable.

However we will see that $T = 0$ will be excluded by other arguments. Note in fact that in (8.33) is not yet introduced the notion of torsion.

A natural way of understanding spin effects in gravitation is through torsion. The modification of the metric by inclusion of torsion can be expressed as:

$$g_{oo} = 1 - \frac{2GM}{Rc^2} + \frac{3G^2s^2}{2R^4c^6} \tag{8.34}$$

The surface gravity of a black hole with torsion can be written as [10]:

$$\chi = \left(-\frac{GM}{R^2} + \frac{3G^2s^2}{c^4R^5}\right)\cdot\text{const.} \tag{8.35}$$

For a mass $M = (\hbar c/G)^{1/2}$ and radius $R = (\hbar G/c^3)^{1/2}$, the Planck length, zero surface gravity would correspond to

$$s^2 = (1/3)(c^4/G)(\hbar c/G)^{1/2}(\hbar G/c^3)^{3/2} = (1/3)\hbar^2 \tag{8.36}$$

implying that for a spin $s \approx \hbar$, for such a black hole, the surface gravity and hence the temperature vanishes. Thus in this case the torsion effects which enter with opposite sign, cancel those of gravity. This gives a minimum radius of $\sim R_{Pl}$.

Writing $s = \sigma R^3$, σ being the spin density, the torsion term in eq.(8.34) would correspond to that of an effective cosmological term of type $\Lambda_{eff} R^2$, the corresponding temperature being of the form

$$\left(\frac{\hbar c^3}{8\pi G k_B M} - \frac{\Lambda_{eff} M G \hbar}{k_B c}\right), \qquad (8.37)$$

vanishing for particular Λ_{eff}, i.e.for a Schwarzschild-de Sitter metric (following the work of Gibbons and Hawking). The entropy of this residual black hole of Planck mass and spin $\hbar$, would be $\sim k_B$, so that the entropy of these black holes'is quantized in units of k_B, just as their spin is quantized in units of $\hbar$. In fact it has been proved that there exists a minimum increase in the areas of black hole surface [11]:

$$(\Delta A)_{min} \geq 8\pi\hbar G/c^3 \qquad (8.38)$$

Combining the well known Bekenstein-Hawking formula we can get

$$(\Delta S\ \ ) = (1/4)\ k_B\ (c^3/G\hbar)\ (\Delta A) \geq 2\pi k_B \qquad (8.39)$$

which means the lower bound of the entropy S increase of black holes is $2\pi k_B$. If the black hole at the final stage of evaporation is of Planck mass, its entropy is also $2\pi k_B$ by Bekenstein-Hawking formula. So it seems reasonable that the entropy of black holes can be quantized in units of k_B.

We will consider in more detail the possibility that the entropy of a black hole is quantized. Now Hawking found the quantum radiation effect of black holes and derived the correct expression of the black hole entropy:

$$S_{bh} \propto c^3 A k_B / G\hbar \tag{8.40}$$

where A is the area surface of theblack hole. Moreover eq.(8.39) indicates that we have a lower bound on entropy increase of black holes in the process that the black holes absorb exterior matter. Inspecting the expression (8.40) of black hole entropy and the expression (8.39) of minimum entropy increase of black holes it is easy to see

$$S_{bh} \propto A/G \propto GM^2\Big|_{M=M_{Pl}} = \hbar c \tag{8.41}$$

$$(\Delta S_{bh})_{min} \propto (\Delta A)_{min}/G = 8\pi\hbar/c^3 \tag{8.42}$$

which are independent of the variation of G and t. So it seems reasonable to propose the conjecture that black hole entropy can be quantized and its quantum unit is $2\pi k_B$.

Indeed the reason why the black hole entropy has a

minimal value ~ k_B for the lowest mass black holes even with variation of G is to be found in the fact that GM^2 is a constant = $\hbar c$ (see eq. (8.11)).

As the black hole entropy is given by

$$S_{bh} \propto \frac{GM^2}{\hbar c} \cdot k_B \qquad (8.43)$$

and as GM^2 = constant = $\hbar c$ for all the fundamental masses (i.e. epoch dependent lowest mass) black holes in the early universe, it follows from (8.43), that S_{bh} is just k_B and cannot be smaller for these black holes. Again eq. (8.43) can be written as

$$S_{bh} \approx k_B (M / M_{Pl})^2. \qquad (8.44)$$

So the entropy for larger mass black holes scales as M^2. For a solar mass black hole ($M \approx 10^{38} M_{Pl}$), $S_{bh} \approx 10^{76} k_B$. For a galactic mass black hole ($M_G \approx 10^{50}\ M_{Pl}$), $S_{bh} \approx 10^{100}\ k_B$ and for a black hole of the mass of the universe ($M_U \approx 10^{60}\ M_{Pl}$), $S_{bh} \approx 10^{120} k_B$. Now, as regards the angular momentum of celestial objects, we have that [12] the angular momentum J of a typical star like the sun is $J_S \approx 10^{76}\hbar$, for a typical galaxy $J_G \approx 10^{100}\hbar$ and for the universe $J_U \approx 10^{120}\hbar$. Now the lowest mass black hole (i.e.remnant of order of Planck mass) has angular momentum which is just $\hbar$ and entropy of just k_B. So for the larger mass black holes, we see that there is a one-to-one correspondence between entropy and angular momentum, i.e. $J \approx 10^{76}\hbar$ implies $S \approx 10^{76}k_B$, $J \approx 10^{120}\hbar$ implies $S \approx 10^{120}k_B$, etc., i.e. every unit of angular momentum $\hbar$ is associated with unit of entropy k_B and as the smallest mass stable black hole must have an angular momentum $\hbar$ (as rotation should support gravitational attraction, as seen above, the lowest unit of black hole entropy is just k_B! The scaling law $S \propto M^2$, is just identical to Brosche law $J \propto M^2$ for angular momentum [13],[14]. In units of $\hbar$, the angular momenta of black holes is $J \approx \hbar\,(G/\hbar c)\,M^2$, identical to the entropy relation $S \approx K_B\,(G/\hbar c)\,M^2$! In both cases

$$J \approx \hbar\,(M/M_{Pl})^2 \quad \text{and} \quad S \approx k_B\,(M/M_{Pl})^2 \qquad (8.45)$$

so that the quantized units of angular momenta and entropy are $\hbar$ and k_B respectively! This further supports the idea of quantized entropy for black holes! Also note that in the context of strong gravity the slope of Regge trajectories is [12]

$$P_h \approx (G_f/\hbar c) \approx (1\mathrm{Gev}/c^2)^{-2}. \qquad (8.46)$$

We have a similar slope of $(G/\hbar c)$ here. Also the spectrum of rotational states of a string has the same slope $p \approx c(2\pi\mu)^{-1}$, where μ is the string tension. This also supports the black hole string transition at the final stage of evaporation.

Again as the energy of a typical quanta emitted by the black hole is ~ $\hbar c^3/GM$, the total number of quanta that can be emitted by a black hole during its lifetime is

$$N \sim \frac{M c^2}{\hbar c^3/GM} \sim \frac{GM^2}{\hbar c} \qquad (8.47)$$

which is the entropy of the hole divided by k_B (i.e. entropy for black-body radiation is the total number of quanta emitted). As the lowest number of quanta is just one, it again follows that the lowest value of the entropy is just k_B.

In 1981 Bekenstein [15] pointed out that there exists a universal upper bound on the entropy-to-energy ratio S/E for an arbitrary system of effective radius R. In c.g.s.units the bound is expressed by:

$$S/E \leq 2\pi R k_B/\hbar c \qquad (8.48)$$

We can give a general justification of the above results, in particular of the upper bound on entropy-to-energy ratio, i.e. that given by eq.(8.48). We start from the definition of entropy as:

$$S = k_B \ln \{\text{number of states}\} \qquad (8.49)$$

which for N particles of the same type would be:

$$S = k_B \ln \{ (\text{number of one-particle states})^N \}$$

$$= k_B \ln \{N^N\} = k_B N \ln N \approx k_B N \qquad (8.50)$$

This gives a simple definition of entropy, i.e. the entropy is just the number of particles in the system multiplied by the Boltzmann constant k_B. For the baryonic matter in the universe, the total number of baryons is $N_B \approx 10^{80}$. So the entropy of baryonic matter is $S_B \approx 10^{80} k_B$. The total energy in the microwave background is

$$2\pi^2 R_H^{\ 3} a\, T^4 \approx 10^{84}\text{ev} \quad (\text{for } T \approx 3\ ^0\text{K},\ R_H \approx 10^{28}\text{cm}) \quad (8.51)$$

As the energy of each photon is $\approx 10^{-4}$ev, the entropy of microwave background is just the total number of photons, i,e, $\approx 10^{88} k_B$. So there is at least 10^8 times more entropy in radiation than in the baryons, i.e. entropy per baryon $\approx 10^8 k_B$. This is important as the entropy of the universe is for all practical purposes given by entropy of relic photons (plus neutrinos and other dark matter) which is much larger than the entropy in the ordered structures such as stars, galaxies etc. This means that the change in entropy required to form galaxies, stars, planets etc. is so small compared to the total entropy that the second law proves no problem for formation of small pockets of order in the universe. From what as been said above, it follows that if E is the total energy of a system of particles and E_{min} the minimum energy of an individual particle, then the maximum entropy would simply be given by:

$$S^{max} \simeq k_B(E/E_{min}) \qquad (8.52)$$

If $E_{min} = \hbar c/\lambda_{max}$, where λ_{max} is the maximum wavelength,

$$S^{max} \simeq k_B(E\lambda_{max}/\hbar c) \qquad (8.53)$$

For black holes, $\lambda_{max} \simeq G M/c^2$, where [7], [10] M is the mass of the black hole. So this would just give (writing $E = Mc^2$) from eq.(8.53)

$$S^{max} \simeq k_B (GM^2/\hbar c) \qquad (8.54)$$

which corresponds with the usual formula for black hole entropy! In general given a system of size R with total (fixed) energy E, we have

$$S^{max} \simeq k_B (ER/\hbar c) \qquad (8.55)$$

For an individual elementary particle, $R = \hbar/mc = \hbar c/E$, so S^{max} is just k_B as implied by the uncertainty principle relation $R = \hbar c/E$. So the entropy associated with each individual particle is just k_B.

6. Torsion, gravity and the Heisenberg uncertainty relations

So through torsion we are able to find some minimal quantity (length, time, entropy, temperature) and then a discretization (quantization) of the space-time, avoiding both the point singularity and the infinities in fields. But we can go a little further.

In fact we can notice [16] that the Landau-Peierls type of uncertainty relations

$$\Delta F(c\Delta t)^2 \geq (\hbar c)^{1/2} \qquad (8.56)$$

(where F denotes the magnitude of the field strength F_{lk} , ΔF its measurement innacuracy, Δt the time interval between two field measurements) corresponding to commutation rules of the form:

$$i\,[F_{\alpha\beta}\,,\, dx^{\alpha\beta} \wedge dx^k] = \sqrt{\hbar c}\,, \qquad (8.57)$$

would imply via Rosenfeld's analysis the uncertainty inequalities for quantum general relativity as:

$$\Delta g_{\alpha\beta}\Delta x^\alpha \Delta x^\beta \geq \hbar G/c^3 \qquad (8.58)$$

and

$$\Delta\Gamma_{\alpha\beta\gamma}\Delta x^\alpha \Delta x^\beta \Delta x^\gamma \geq \hbar G/c^3 \qquad (8.59)$$

(L_o denotes the dimension of the spatial region over which the average value of $g_{\alpha\beta}$ and $\Gamma^{\alpha}{}_{\beta\gamma}$ is measured and Δg and $\Delta\Gamma$ the corresponding innacuracy) However in general relativity the "field strength" $\Gamma^{\alpha}_{\beta\gamma}$ (which are the usual Christoffel symbols) are not tensorial quantities. So distances cannot be defined independently of the field quantities $g_{\alpha\beta}$. In fact one should not expect to find commutation rules corresponding to Rosenfeld's inequality relations (8.58) and (8.59) because the quantities "field strength" and "length" appearing in (8.58) and (8.59) cannot be defined independently of each other. The independent definition of those quantities for which commutation rules are to be formed is however necessarily to require. Therefore, the inequalities (8.58) and (8.59) do not have the status of uncertainty relations like (8.56) and (8.57); they are rather relations describing inaccuracy for the measurement of the single quantities $g_{\alpha\beta}$, $\Gamma^{\alpha}{}_{\beta\gamma}$, etc. As a consequence of the dependence of distances on the metric $g_{\alpha\beta}$, the inequalities (8.58) and (8.59) say of course also something about limitations on distance measurements. Indeed, assuming e.g. a coordinate system in which $|g_{oo}| = 1$ is satisfied, with the condition $|\Delta g_{oo}| < |g_{oo}|$ (accoording to Rosenfeld $|\Delta g| < |g|$ is the most modest requirement because the inaccuracy of a quantity has of course to be smaller than the value of the quantity itself) one is led from (8.58) to

$$L_o^2 > |\Delta g_{oo}| L_o^2 \geq \hbar G/c^3 \qquad (8.60)$$

showing that for distances of the order of magnitude of Planck's length the notion "length" loses its physical meaning. But this statement about limitations on field and length measurements does not coincide with the statement implied by uncertainty relations of type (8.56) and valid for canonically conjugate quantities.

Therefore, the conclusion is that the inequalities (8.58) anfd (8.59) establish limitations on the validity of the relativistic quantum field conception. And it was stressed that such limitations should only occur in quantized General Relativity Theory but not in any other theory of quantum gravity. Such limitations should not arise in a gravitational theory that does not satisfy the strong principle of equivalence [17 18]. In ref.[18] it was shown that even in electrodynamics limitations occured if one would postulate an equivalence principle.

A weak version of such a postulate can be introduced into electrodynamics by requiring that the specific charge Q/M of all bodies and particles (i.e. of all the electrically charged matter) is equal to the specific charge of electrons

$$Q/M = e/m \tag{8.61}$$

If one does so, then one can or even must accept Dirac's (of course Dirac's theory is wrong because electromagnetic interaction does not satisfy an equivalence principle; but here we consider it since we ask: "what happened if such principle would be valid?) electrodynamics where the vector potential A_α is given by

$$A_\alpha = (mc/e)(dx^\alpha/dt) = (c/e)p_\alpha \qquad (8.62)$$

Equation (8.62) leads via Heisenberg's uncertainty relations to

$$\Delta A_\nu \; \Delta x^\mu \geq \hbar c/e\delta^\mu_\nu \quad (\mu,\nu = 1,2,3), \qquad (8.63)$$

$$\Delta A_0 \; \Delta t \geq \hbar/c \qquad (8.64)$$

The derivation of relations (8.63) and (8.64) shows the close relation between the equivalence principle and relations like (8.56), (8.58) and (8.59).

In contrast to the inequality relations (8.58),(8.59) valid in General Relativity Theory, the inequality relations (8.63),(8.64) are however genuine uncertainty relations corresponding to commutation rules, namely to

$$i[A_\beta \, , \, \Delta x^\alpha] = (\hbar c/e)\delta^\alpha_\beta \qquad (8.65)$$

This is due to the fact that in Dirac's electrodynamics only the weak equivalence principle (8.61) is satisfied, whilst in General Relativity Theory not only the universal constancy of the specific gravitational) charge is valid, but, moreover, metric and (gravitational) fieldsare identified. This identification establishes a strong version of the equivalence principle and leads to the problem as to whether there are commutation rules corresponding to (8.58),(8.59).

This problem may not arise if torsion is considered, since in this case the asymmetric part of the connection i.e. $\Gamma^{\alpha}_{[\beta\gamma]}$, i.e.the torsion tensor $Q^{\alpha}{}_{\beta\gamma}$ is a true tensorial quantity. In fact with torsion one can define distances in the following sense: we consider a small closed circuit and write $l^{\alpha} = \oint Q^{\alpha}{}_{\beta\gamma} dA^{\beta\gamma}$ where $dA^{\beta\gamma} = dx^{\beta} \wedge dx^{\gamma}$ is the area element enclosed by the loop. Then l^{i}represents the so called "closure failure", i.e. torsion has an intrinsic geometric meaning: it represents the failure of the loop to close, l^{α} having the dimension of length. ($Q^{\alpha}{}_{\beta\gamma}$ has dimension of inverse length, $dA^{\beta\gamma}$ is area).

For in the geometrical description of crystal dislocations or defect it is known that torsion plays the role of defect density, i.e. we now treat space-time as a deformable medium with elastic properties in the spirit of Sakharov. In the above relation, torsion can be related to the intrinsic spin $\hbar$, by postulating that the defect in space time topology should occur in multiples of the Planck length, $(\hbar G/c^{3})^{1/2}$ i.e.

$$\oint Q^{\alpha}{}_{\beta\gamma} dx^{\beta} \wedge dx^{\gamma} = n\ (\hbar G/c^{3})^{1/2} \qquad (8.66)$$

where n is an integer. This is analogous to the well known $\oint pdq = n\hbar$, i.e. the Bohr-Sommerfeld relation. As $Q^{\alpha}{}_{\beta\gamma}$ plays the role of the field strength (analogous to $F_{\alpha\beta}$ for electromagnetism), relation (8.66) is analogous to that of eq.(8.57) (with $\sqrt{G\hbar}$ on the right hand side). So distance has been defined independently of $g_{\alpha\beta}$. In fact (8.66) would define a minimal fundamental length, i.e. the Planck length entering through the minimal unit of spin or action $\hbar$. So $\hbar$ has to deal with the intrinsic defect built into torsion structure of space time through [19] $l^{\alpha} = \oint Q^{\alpha}_{\beta\gamma} dx^{\beta} \wedge dx^{\gamma}$, i.e. we have

$$[Q^{\alpha}_{\beta\gamma} \; , \; dx^{\beta} \wedge dx^{\gamma}] \geq (\hbar G/c^3)^{1/2} \qquad (8.67)$$

i.e. $\hbar$ is related to a quantized timelike vector with dimensions of length related to an intrinsic geometrical structure with torsion.

Regarding that the intrinsic spin $\hbar$ is a universal attribute of all matter, torsion must even enter into geometry.

In the context of the SL(2,C) gauge invariant formalism of the Einstein-Cartan theory, it has been shown [4], [19] that the Einstein-Cartan equations may be expressed in the form of elegant commutation relations of the type:

$$i\,[\gamma \;, \Gamma_{\mu\nu}] = \chi\, J_{\nu} \qquad (8.68)$$

$$\nabla_{\mu}\, [\gamma^{\mu}, \gamma^{\nu}] = \chi\, S^{\nu} \qquad (8.69)$$

in terms of a conserved set of SL(2,C) Noether currents

$$J^{\nu} = S^{\nu} - i\left[\Gamma_{\mu}, [\gamma^{\mu},\gamma^{\nu}]\right] \qquad (8.70)$$

which are identically conserved

$$\partial_{\mu}J^{\mu} = \partial_{\nu}\partial_{\mu}[\gamma^{\mu},\gamma^{\nu}] = 0 \qquad (8.71)$$

All these quantities are tensorial quantities, and the formalism is now analogous to electromagnetism or Yang Mills theory with the gauge group now being SL(2,C) instead of U(1) or SU(2). Torsion is necessary for these invariances. $\Gamma^{\mu\nu}$ is the spin curvature, analogous to the field strength $F_{\mu\nu}$. γ^{μ} is the space-time dependent Dirac matrix. The metric is defined through the γ's as $\gamma^{\mu}\gamma^{\nu} + \gamma^{\nu}\gamma^{\mu} = 2g^{\mu\nu}$. Quantization of the spinor current in unit of $\hbar$ would then give the desired uncertainty relation, γ^{μ} (defining metric) and $\Gamma_{\mu\nu}$, in eq. (8.68), playing the role of conjugate coordinate and momenta, analogous to $[x,p] = i\hbar$. J^{ν} quantized in units of $(\hbar c)^{1/2}$.

Again the equation of geodesic deviation of a free particle with non-zero spin (in which coupling of spin to the space-time curvature occurs), is of the form:

$$\frac{d^2x^{\mu}}{ds^2} + \left\{ {\mu \atop \alpha\beta} \right\} \frac{dx^{\alpha}}{ds}\frac{dx^{\beta}}{ds} = Q^{\mu}_{\alpha\beta}\frac{dx^{\alpha}}{ds}\frac{dx^{\beta}}{ds} \qquad (8.72)$$

In the quantum picture, the right hand side following the discussion above would be in multiples of the Planck length. As the left hand side of eq.(8.72) is obtained from variation of $\int g_{\alpha\beta}(dx^{\alpha}/ds)(dx^{\beta}/ds)$, we would simply have

$$\Delta\Gamma^{\mu}_{\alpha\beta}\, dx^{\alpha}\wedge dx^{\beta} = [Q^{\mu}_{\alpha\beta}\, , dx^{\alpha}\wedge dx^{\beta}] \geq (\hbar G/c^{3})^{1/2} \qquad (8.73)$$

In the limit of vanishing torsion, we have the usual geodesic equation. So torsion in eq.(8.72) can be identified with the quantum correction (in multiples of $\hbar$) to the classical equation of motion.

Again, unlike electromagnetism where a single particle suffices to measure the components of the field tensor, in the case of gravitation we require the relative acceleration between two particles to specify the components of the curvature tensor, i.e. $R^{\alpha}{}_{\beta\gamma\delta}$ is defined through the acceleration:

$$d^{2}x^{\alpha}/ds^{2} = R^{\alpha}{}_{\beta\gamma\delta}(dx^{\beta}/ds)(dx^{\gamma}/ds)n^{\delta} \qquad (8.74)$$

(n^{δ} = unit vector in direction of the line joining the two particles).

Many approaches [20],[21],[22] led to the notion of "maximal acceleration"; for instance in the case of gravitation, this acceleration amounts to $\sim m_{Pl}c^{3}/\hbar = c^{7/2}/(\hbar G)^{1/2}$ (m_{Pl} = Planck mass) [23]. This maximal value for d^2x^α/ds^2 would also imply a maximal curvature which turns out to be $\approx c^3/\hbar G \approx 10^{66}\text{cm}^{-2}$ [24]. In the classical limit, i.e. $\hbar \Longrightarrow 0$, we have the usual curvature singularity, i.e. the maximal value is infinite, the relation between acceleration and curvature being of the form: $a = c^2\sqrt{R}$, where R is the curvature scalar. This would imply a relation of the type

$$\Delta R_{\alpha\beta\gamma\delta}(\Delta x^\alpha \wedge \Delta x^\beta)(\Delta x^\gamma \wedge \Delta x^\delta) \geq \hbar G/c^3. \qquad (8.75)$$

So a general relativistic upper limit on acceleration due to quantum effects (an operational way of measuring curvature tensor) would also imply a maximal curvature and hence uncertainty relation involving Planck length.

Thus in order to arrive at a quantum theory of gravitation for which no limitations but only genuine uncertainty relations occur one has to decouple metrical background and quantized (gravitational) field, i.e. to modify General Relativity Theory (see [19]). The gravitational theories with torsion discussed here are of this nature. There one has a classical background metric $g_{\alpha\beta}$ (saying, for instance, what $dx^\alpha \wedge dx^\beta$ means) and an (nearly) independent tensor field $Q^\alpha{}_{\beta\gamma}$ quantized via condition (8.66):

$$\oint Q^\alpha{}_{\beta\gamma} dx^\beta \wedge dx^\gamma = n\ (\hbar G/c^3)^{1/2} \qquad (8.76)$$

this quantization leads to the commutation rules (8.67):

$$[\, Q^{\alpha}{}_{\beta\gamma} \,,\, dx^{\beta} \wedge dx^{\gamma} \,] \quad = \quad (\hbar G/c^{3})^{1/2} \qquad (8.77)$$

and to corresponding uncertainty relations:

$$\Delta Q^{\alpha}{}_{\beta\gamma} \; (\Delta x^{\beta} \wedge \Delta x^{\gamma}) \;\geq\; (\hbar G/c^{3})^{1/2} \qquad (8.78)$$

Therefore, the Einstein-Cartan theory of gravitation should in contrast to Einstein's General Relativity Theory, provide genuine quantum-gravity effects.

If we wish to connect the initial and final positions of one and the same particle we cannot avoid uncertainty associated with torsion, i.e. for a sufficiently small area element, dS, the uncertainty in position vector between initial and final position would be $\Delta l^{\mu} = Q^{\mu} dS$, and this would induce fluctuations in distance in defined the metric through $\Delta l = (g_{\mu\nu} dx^{\mu} dx^{\nu})^{1/2}$. So what is important is not the points themselves but the 'fluctuations' in their position i.e. the interval between them caused by a plastic deformation of space itself through torsion. Note that plastic deformations are induced by torsion and are different from the elastic deformations considered by Sacharov (which depend only on curvature). With quantized values, these fluctuations would manifest as metric fluctuations. Since curvature causes relative acceleration between neighbouring points we have momentum uncertainty related to curvature as:

$$ma^{\mu}ds = \Delta p^{\mu} = m\ R^{\mu}{}_{\alpha\beta\gamma}(dx^{\alpha}/ds)\ dx^{\beta}n^{\gamma} = m\ R\ (dS/ds)n^{\mu} \qquad (8.79)$$

So as position fluctuations are caused by torsion, momentum fluctuations are due to curvature and we can interpret quantum effects (and thus the uncertainty principle) geometrically as consequences of space-time deformation, i.e. [25]:

$$\Delta p^{\mu}\Delta x_{\mu} = \hbar \qquad (8.80)$$

where

$$\Delta x^{\mu} = Q\ c\ ds\ n^{\mu}, \quad \text{and} \quad \Delta p^{\mu} = m\ R\ (dS/ds)n^{\mu} \qquad (8.81)$$

(i.e. uncertainty in initial and final positions and relative acceleration between them due to space-time deformations), Q (torsion) and curvature (R) thus both playing simultaneous roles as conjugate variables of the geometry (gravitational field), thus enabling us to write commutation relations between curvature and torsion as (analogous to $[x,p] = i\hbar$) as:

$$\left[Q\ ,\ R\right] = (\hbar G/c^{3})^{-3/2} \qquad (8.82)$$

or, more explicity, as:

$$\left[Q^{\alpha}{}_{\mu\nu}\, dx^{\mu} \wedge dx^{\nu} \;,\; R^{\alpha}{}_{\mu\nu\rho}\, dx^{\mu} \wedge dx^{\rho} \wedge n^{\nu}\right] = i\hbar G/c^{3} \qquad (8.83)$$

or

$$\Delta Q\, \Delta R \geq L_{Pl}^{-3} \qquad (8.84)$$

where L_{Pl} is the Planck length.

Thus in a sense, we have written quantum commutation relations for the observables of the background geometry, rather than for the gravitational field in a fixed background as in the usual picture; thus partially removing the inconsistency between quantizing the gravitational field and not the geometry, gravity being the geometry itself!

In fact in the usual methods of quantizing the gravitational field one has the field on a fixed background, which gives rise to an inconsistency, because the gravitational field is just geometry so that how can we separate the geometry and the field? Here, on the contrary, we have improved our previous work where we had quantized torsion on a fixed background but had not connected torsion with curvature. In fact here we have considered both torsion and background curvature which appear as conjugate variables.

The existence of a minimal length and time, l_{min} , t_{min} has important consequences for quantum mechanics, at very short distances (high energies!). It implies for instance uncertainty relations even for coordinate measurements, independent of momenta.

In additions to commutation relations for angular momentum operator (owing to existence of a minimal action $\hbar$)

$$[L_x , L_y] = i \hbar L_z ; \quad [L_y , L_z] = i \hbar L_x \quad \text{etc.} \quad (8.85)$$

we have

$$[x , y] = i\, l_{min} z ; \quad [y , z] = i\, l_{min} x \quad \text{etc.} \quad (8.86)$$

Therefore in addition to the usual uncertainty relations

$\Delta p\, \Delta x \geq \hbar/2$, etc.

we have also

$$\Delta x\, \Delta y \geq (1/2) l_{min} \Delta z ,$$

$$\Delta x\, \Delta z \geq (1/2) l_{min} \Delta y , \quad \Delta x\, \Delta t \geq (1/2) t_{min} \Delta x \quad \text{etc.}$$

This raises the interesting possibility that space-time itself at small scales is non-commutative.

Another feature of non-commutative space-time is the existence of uncertainty relations between momentum components. This is similar to what happens for Landau levels, where motion transverse to the megnetic field is quantized i.e.

$$[p_x , p_y] = (ie\hbar/mc) B_z , \text{ giving } \Delta p_x \Delta p_y = (1/2)(e\hbar/mc) \Delta B_z \quad (8.87)$$

As torsion also generate a magnetic field, we can relate B to Q as:

$$B \Longrightarrow 2c^2Q(\alpha/G)^{1/2} \qquad (8.88)$$

This means that torsion at small scales would lead to uncertainty relations between momentum components giving rise to uncertainty relations between coordinates.

7. Torsion and twistors

The so called twistor formalism introduced by Penrose [26] provides a different view of quantized space-time. In the usual view one imagines a space-time where points remain intact bur g_{ab} becomes quantized.

In the twistor view, the space-time 'direction of motion' of twistors (a twistor representing a massless particle in motion) remains null, the points of space-time intersection between twistors become subject to quantization, i.e. the points become 'fuzzy' or fluctuate quantum mechanically.

We straigthway notice the striking analogy to our notion of line defects induced by torsion. Also in our picture space-time points are not fundamental, but the line defects are, and intersections between defects gives rise to the space-time points whose separation is now quantized in terms of the non-closure property due to torsion as $\oint Q^{\gamma}_{\alpha\beta}\, dx^{\alpha}dx^{\beta} \geq$

$(\hbar G/c^3)^{1/2}$. Fluctuations in position due to torsion and in momentum due to curvature enabled us to write quantum uncertainty relations for the conjugate variables of curvature and torsion (see previous section).

In the twistor picture to fix a space-time point P, one considers trajectory or path of massless particles in free motion through P (i.e. twistor) and it is this timelike or null vector at P that becomes subject to quantization.

In torsion case, as seen in previous section, the spin $\hbar$ is related to a timelike vector through the torsion and the resulting non closure is subject to quantization. The points are derived or secondary concepts in both cases. In both cases the spinors are more fundamental than the world vectors.

In the twistors case, the physical world is pictured as a sort of network of line segments, each segment is thought of as an 'object' having a well defined angular momentum of $n\hbar/2$. Intersections of segments give rise to space-time points, these points then being derived objects, the twistors themselves being more basic.

In the torsion case we have a network of line defects, each associated with an intrinsic spin of $n\hbar/2$.

In the twistor case complex spin vectors (on which SL(2,C) invariant elements act when a local Lorentz transformation is performed) are regarded as primary. All world vectors and tensors are then taken to be built up of spin-vectors by means of tensor products, sums etc.

In the torsion picture as we have seen, the spinor gauge field (described by tetrads) which transform invariantly under SL(2.C) gauge invariance (under local Lorentz rotations of tetrad) are the basic entity. In fact the metric is a derived concept, at every point (described by intersection of defects) it is the expectation value of a tetrad product, i.e.

$$\langle h_a^b \, h_c^d \, \eta_{bd} \rangle = g_{ac} \qquad (8.89)$$

Curvature tnsors and other geometric quantities are built up out of these tetrads as seen in Chapter 5, in connection with the Dirac equation.

Analogously in the twistor case the metric is defined as

$$g_{ab} = \langle \varepsilon_{AB} \, \varepsilon_{A'B'} \rangle \qquad (8.90)$$

in terms of complex spin vectors.

The motivation for using complex numbers to define the structure of twistor space, has to do with the <u>complex</u> Hilbert space of quantum modes, on the one hand, while on the other hand we have probability amplitudes superposition laws, etc. in <u>real</u> space-time. Usually these two continua are conventionally regarded as independent of each other. One of the basic features of twistor theory is that at the very outset, these two continua are attempted to merge into one, that is, the complex quantum mechanical continuum becomes woven into space-time geometry. Thus we have an interrelation between geometry and complex numbers (entities like probability amplitudes need complex numbers!). Moreover in twistor theory rest mass is regarded as a secondary concept. Also in the torsion picture it is secondary, the quantized vortices associated with torsion are more basic and they lead to rest mass as we have seen earlier.

There are many ways to formally introduce twistor concepts. The most directly physical one, relevant for our purpose, is to use angular momentum of zero rest mass particle. The total momentum 4-vector $\bar{p}_a$ and angular momentum bivector $M^{ab} = - M^{ba}$, with respect to some origin O correspond to those of a zero rest mass particle of helicity h. We have

$$\bar{S}_a = h\,\bar{p}_a \tag{8.91}$$

$\bar{S}_a$ being the Pauli-Lubanski vector, being given by

$$\bar{S}_a = (1/2)\varepsilon_{abcd}\bar{p}^b M^{cd} \tag{8.92}$$

$\bar{p}^a$ is nulland future pointing i.e. $p_a p^a = 0$ and $p^0 > 0$.
The two-spinor expression $p_{AA'}$ for p^a has form

$$p_{AA'} = \pi_A\ \pi_{A'} \tag{8.93}$$

(for some dual conjugate spin vector π_A).
The angular momentum in this notation is defined as [27]

$$M^{AA'BB'} = i\,\omega^{(A}\,\bar{\pi}^{B)}\varepsilon^{A'B'} - i\,\varepsilon^{AB}\,\bar{\omega}^{(A'}\pi^{B')} \qquad (8.94)$$

The conjugate pair (ω^A, π_A) uniquely defines momentum p_a and angular momentum M^{ab}.

The gauge freedom in ω^A , π_A is of overall phase change, i.e. invariance under

$$\omega^{A'} \Longrightarrow e^{i\vartheta}\,\omega^A\,, \quad \pi_{A'} \Longrightarrow e^{i\vartheta}\pi_A \quad (8.95)$$

ω^A , π_A are 2 spinor parts of twistors Z^α, written as:

$$Z^\alpha \Longleftrightarrow (\omega^A\,,\ \pi_A) \qquad (8.96)$$

The complex conjugate of twistor Z^a , is the dual $\bar{Z}_\alpha$, i.e.

$$\bar{Z}^\alpha \Longleftrightarrow (\bar{\pi}_{A'}\,,\ \bar{\omega}^{A'}) \qquad (8.97)$$

The scalar product $Z^a\bar{Z}_\alpha$ can be now formed as:

$$Z^\alpha\bar{Z}_\alpha = \omega^A\,\bar{\pi}_A + \pi_{A'}\bar{\omega}^{A'} \qquad (8.98)$$

This so called twistor norm, $Z^\alpha\bar{Z}_\alpha$, physically measures the helicity of a zero mass particle i.e.:

$$Z^{\alpha}\bar{Z}_{\alpha} = 2 h \qquad (8.99)$$

We can think of a complexified space-time co-ordinate as:

$$Z_{\mu} = x_{\mu} \pm i h_{\mu} \qquad (8.100)$$

where $i h_{\mu}$ corresponds to the 'internal helicity' projection vector attached to the space-time point x_{μ} .

Space-time point can be written as a 2 **x** 2 complex matrix having SL(2,C) group structure (physically two opposite orientations of project vector or internal helicity ($h p_a$) can correspond to particle and antiparticle in isospin space).

Considering a complex manifold, the coordinates in this extended space can be written

$$Z^{\mu} = x^{\mu} + i \eta^{\mu}_{a} \vartheta^{a} \qquad (8.101)$$

or, in matrix representation:

$$Z^{BB'} = x^{BB'} + i \eta^{BB'}_{a} \vartheta^{a} \qquad (8.102)$$

($\Theta(\bar{\vartheta})$ is a spinorial direction vector attached to a space time point)

$$x^{BB'} \qquad \begin{bmatrix} x^0 - x' & x^2 - ix^3 \\ x^2 - ix^3 & x^0 - x' \end{bmatrix} \qquad (8.103)$$

$$\eta_a^{BB'} \in SL(2,C) \qquad (8.104)$$

The twistor equation in terms of the complex conjugate spinors $\bar{\pi}_B$ $(\pi_{B'})$ is:

$$\bar{Z}_\alpha Z^\alpha + \eta_a^{BB'} \vartheta^\alpha \bar{\pi}_B \pi_{B'} = 0 \qquad (8.105)$$

The four momentum corresponds to:

$$\bar{p}^{BB'} = \bar{\pi}^B \pi^{B'} \qquad (8.106)$$

$$Z^\alpha = (\omega^B , \pi_{B'}) ; \qquad \bar{Z}^\alpha = (\bar{\pi}_B , \omega^{B'}) \qquad (8.107)$$

The helicity operator is then:

$$h = Z^\alpha \bar{Z}_\alpha = - \eta_\alpha^{BB'} \vartheta^\alpha \bar{\pi}_B \pi_{B'} \qquad (8.108)$$

$\bar{p}^\mu$ is the canonical conjugate of x^μ in $Z^\mu = x^\mu + i\eta^\mu$.

$$\bar{p}^{BB'} = \bar{\pi}^B \pi^{B'} , \qquad \bar{p}_\mu^2 = 0 \qquad (8.109)$$

Defining

$$\beta = i \vartheta^{A'} \pi_A , \qquad \bar{\beta} = - i \bar{\vartheta}^A \bar{\pi}_{A'} \qquad (8.110)$$

$$S = - \vartheta^A \vartheta^{A'} \bar{\pi}_A \pi_{A'} = \beta \bar{\beta} \qquad (8.111)$$

The metric is then:

$$g_{\mu\nu}(x, \vartheta, \bar{\vartheta}) = g_{\mu\nu}^{BB'}(x) \bar{\vartheta}^B \vartheta^{B'} \qquad (8.112)$$

One can now construct connections (both symmetric and antisymmetric) from this metric and its derivatives. Torsion and curvature corresponding to this complexified space can in turn be constructed from these connections i.e.

$$D\vartheta^{i} = (1/2)Q^{i}_{jk}\ \vartheta^{j} \wedge \vartheta^{k}\ , \quad D\ \omega^{i}_{j} = (1/2)R^{i}_{\ jkl}\vartheta^{j} \wedge \vartheta^{k} \qquad (8.113)$$

One can define radius vector of field which can be integrated along a curve and when this is done for a loop (of radius r), one finds that with torsion, it does not return to initial value, i.e. it gives rise to a local translation in the internal space of the radius vector field. So the change in the intersection loop of two twistors when taken along such a loop change is given by:

$$\delta\ Z^{\alpha} = \frac{1}{r^{2}} \int (d\omega^{B}_{ji} + \omega^{B}_{ki} \wedge \omega^{Bk}_{j}) Z^{\alpha}\ dx^{i}dx^{j} \geq \hbar^{1/2}n^{\alpha} \qquad (8.114)$$

The fluctuation in the norm $\langle \delta Z^{\alpha}\ \delta\bar{Z}_{\alpha} \rangle \geq \hbar$ then gives a twistor geometric definition of Planck's constant. Thus this automatically leads to the quantization of the background geometry. This may provide a basis for the twistor commutation relation:

$$\left[Z^{\alpha}\ ,\ Z^{\beta}\right] = 0, \quad \left[\bar{Z}_{\alpha}\ ,\ \bar{Z}_{\beta}\right] = 0\ , \quad \left[Z^{\alpha}\ ,\ \bar{Z}_{\beta}\right] \geq \hbar\ \delta^{\alpha}_{\beta} \qquad (8.115)$$

Another attractive feature of this approach is an automatic explanation of the dimensionality and signature of space-time. Wee had seen the necessity (also in chap.V) while discussing the gauge invariance of the theory of using the group SL(2,C) (this incorporates spin into the structure of space-time through torsion giving a Yang-Mills type of description). This group is also the one relevant for the twistor description of space-time (where again the basic geometric entity has spin). Now SL(2,C) is isomorphic to O(3,1) which is the rotation group in 3 + 1 dimensions, i.e. 3 space and one time (with signature - + + + or + - - -).[*] So all the interesting physical consequences we had earlier, would imply 4 dimensions with 3 spatial and one temporal dimension.

Thus the signature and dimensionality of space time is naturally contained in the structure of the theory, and we do not have to invoke ad hoc hypothesis or even impose the antropic principle as has been done by some authors.

---

[*]We note that the finitess of the velocity of light 'c' implies a light cone structure giving opposite signature to time and space. In addition the presence of the quantum of spin ħ, imposes the dimensionality through torsion, as for the complete gauge invariant description of the theory we need four and only four dimension.

References

[1] V.de Sabbata and C.Sivaram - Ann.der Physik 47, 511 (1990)

[2] V.de Sabbata and C.Sivaram - Il Nuovo Cimento 105B, 1181 (1990)

[3] V.de Sabbata and C.Sivaram - Il Nuovo Cimento 105B, 603 (1990)

[4] V.de Sabbata and C.Sivaram - "Torsion and Quantum effects" in 'Modern Problems of Theoretical Physics' (Festschrift for Professor D.Ivanenko) ed.by P.I.Pronin and Yu.N.Obukhov World Sci.Singapore 1991 pg.143

[5] V.de Sabbata and C.Sivaram - Foundations of Phys.Letters 6, 201 (1993)

[6] P.A.M.Dirac - "New directions in Physics" J.Wiley N.Y.(1978)

[7] V.de Sabbata and C.Sivaram - "Torsion, Quantum Effects and the problem of cosmological constant" in vol. devoted to 75th jubilee of Peter Gabriel Bergmann - ed.by Antonino Zichichi, Venzo de Sabbata and Norma Sanchez, Plenum Press E.Majorana Int.Sci.Series Vol.56, N.Y.1991 pg.19

[8] S.W.Hawking - Commun.Math.Phys.43, 199 (1975); Nature 246, 30 (1974)

[9] N.Birrel and P.Davies - "Quantum fields in curved space" CUP, 1962

[10] V.de Sabbata and C.Sivaram - 'The final state of an evaporating black hole'in "Quantum Mechanics in curved space-time" ed.by J.Audretsch and V.de Sabbata, Plenum Publ. Corp.N.Y.,1991, pag.485; see also V.de Sabbata, C.Sivaram and D.Wang - Ann.der Physik 47, 508 (1990)

[11] J.D.Bekenstein - Phys.Rev.D7, 2333 (1986) (1989)

[12] V.de Sabbata and C.Sivaram - Il Nuovo Cimento 100A, 919,(1988)

[13] P.Brosche - in "Spin, Torsion, Rotation and Supergravity" ed.by Peter G.Bergmann and Venzo de Sabbata, Plenum N.Y. 1980 pg.375

[14] V.de Sabbata - 'On the relations between fundamental constants' in "Gravitational Measurements, Fundamental Metrology and Constants" ed.by Venzo de Sabbata and V.N.Melnikov, Kluver Academic Publ.1988 pg.115

[15] J.D.Bekenstein - Phys.Rev. D23, 287 (1981)

[16] H.-H.v.Borzeszkowski and H.-J.Treder - Ann.der Physik 46, 315 (1989)

[17] H.-H.v.Borzeszkowski and H.-J.Treder - "On Quantum Gravity" PRE-EL 88-05, Potsdam 1988.

[18] H.-H.v.Borzeszkowski and H.-J.Treder, The Meaning of Quantum gravity, Dordrecht, Reidel Publ.Comp. 1988.

[19] V.de Sabbata, C.Sivaram, H.-H.v.Borzeszkowski and H.-J.Treder Ann.der Phys.(Leipzig) 48, 197 (1991)

[20] E.R.Caianiello - Nuovo Cimento Letters 32, 65 (1981)

[21] E.R.Caianiello - Nuovo Cimento Letters 41, 370 (1984)

[22] M.Gasperini - Astrophys.Space Sci.135, 387 (1987)

[23] V.de Sabbata and C.Sivaram - Astrophys.Space Sci. 176, 145 (1991)

[24] V.de Sabbata and C.Sivaram - "On limiting Field Strengths in Gravitation", Found.Phys.Letters, Dec.1993.

[25] V.de Sabbata and C.Sivaram - Found.Phys.Letters 5, 579 (1992)

[26] R.Penrose - in "Cosmology and Gravitation, Spin, Torsion, Rotation and Supergravity" ed.by Peter G.Bergmann and Venzo de Sabbata, Plenum Press N.Y.Series B: Physics Vol.58, 287 (1980)

[27] R.Penrose - in "Battelle Rencontres 1967" ed.by Cécile M. DeWitt and John A.Wheeler, Benjamin,N.Y. 1968 pg.121.

CHAPTER IX

NEGATIVE MASS AND TORSION

1. Introduction

Though there is no empirical evidence for the existence of negative mass there is no a priori justification for entirely excluding it from framework of theoretical physics. For instance as noticed early by Terletsky [1] and others [2], in the structure of both the special and general theories of relativity it is possible to introduce positive as well as negative masses. This follow from the definition of rest or proper mass in special relativity given by (units of c = 1):

$$M_o^2 \;=\; E^2 - p^2 \;, \text{ so that } \; M_o \;=\; \pm(E^2 - p^2)^{1/2} \qquad (9.1)$$

The mass-energy equivalence principle would then imply that a particle of negative proper mass, would also have a negative rest energy $E_o = - M_o c^2$ (c of course is positive!) and a negative kinetic energy of $K = - (m_o c^2)/(1-v^2/c^2)^{1/2}$, at ultrarelativistic energies, $K - mc^2$; $-m = -m_o/(1-v^2/c^2)^{1/2}$ so that as $v \Longrightarrow c$, $m \Longrightarrow -\infty$, in contrast to that for positive proper mass particles, where $m \Longrightarrow +\infty$ as $v \Longrightarrow c$.

In general relativity this would mean that the sign of the temporal component (T_{oo}) of the energy-momentum tensor can be either positive or negative, the field equations $G_{ik} = |\chi T_{ik}|$, remaining the same. Of course the gravitational properties of negative mass particles (henceforth called negatons, as suggested by Terletsky), assuming they exist, would be very interesting and in contrast to that of the usual positive mass particles dubbed positons.

For instance two negative mass particles would gravitationally repel one another so that a collection or assembly of such particles must distribute themselves in the universe almost homogeneously. Here again we emphasize that the assumption of the strict validity of the equivalence principle of general relativity would imply that a negative inertial mass is tantamount to a negative gravitational mass.

This point must be stressed because some authors [3] have raised the question of whether antiparticles such as positrons and antiprotons have a negative gravitational mass. We know that they have positive inertial mass, because when they annihilate we get quanta of energy at least equal to their combined rest energy! The CPT theorem would tell us that their inertial masses should be strictly equal.

If the CPT theorem is valid for gravity, we would expect their gravitational masses to be identical as well but some authors still argue that their gravitational masses could have opposite sign. In fact experiments are under way [4] with the low energy antiproton storage ring to see whether antiprotons would behave the same way as protons in the Earth's gravitational field.

An early argument due to Schiff [5] would refute the idea that positrons have negative gravitational mass. He argued that virtual electron-positron pairs surrounding nuclei give rise to the Uehling [6] shift (~ 27 Mc for hydrogen) and this depends on the magnitude of the nuclear charge (i.e. Z). So if positrons had a negative gravitational mass, one would expect anomalies to show up in the Eötvos experiment which shows no deviation to one part in 10^9 between materials with a wide range in Z, eg. aluminium and lead (another argument against negative mass of antiparticles is given by Winterberg [7]).

As far as antineutrinos are concerned, they would seem to have the same gravitational interaction as neutrinos to one part in 10^6 as shown by the SN1987a detection of neutrinos and antineutrinos [8]. Nevertheless a direct experimental test to see whether the gravitational mass of antiparticles is anomalous is soon to be performed. Supergravity and superstring models would predict such differences in gravitational properties of particles and antiparticles but not in the sign of the mass. However the validity of the equivalence principle would compel us to assign a <u>positive</u> gravitational mass to antiparticle as their inertial mass is known to be positive. But in the case of test particles with a <u>negative</u> mass (i.e. by above argument both their inertial and gravitational masses have to be <u>negative</u>), they would <u>fall</u> in the field of a positive mass source such as the Earth. This is because we have $-ma = GM(-m)/r^2$ with $(-m)$ cancelling out (again because of the equivalence principle).

So all test particles whether of positive or negative mass should fall in the field of a positive mass source.

A negative mass particle would move in the opposite direction to that of the force applied on it in general. So we can make a general statement that a positive mass source 'attracts' both positons and negatons, while a negative mass source 'repels' everything! Suppose we have a body of mass M: a small particle of mass Δm is slowly brought from infinity to the surface of M. Then the net (or total mass) including potential energy is:

$$M + \Delta M - M(\Delta M)G/rc^2 = M' \tag{9.2}$$

If both M and ΔM (Case I) are +ve then for a radius $r \approx GM/c^2$, $M' = M$, i.e. the test particle confers no extra mass; if $r < GM/c^2$, M' can be reduced and in fact is $< M$. While this is forbidden in general relativity (as radius cannot be smaller than gravitational radius), in Newtonian theory it is possible to collapse the body to a point so that M' can become $-\infty$! thus extracting infinite amount of energy. In general relativity the maximum energy that can be extracted is just M (in the more general case it is referred to as Buchdalh's theorem).

Now (Case II) if M is positive and ΔM negative we have

$$M - \Delta M + M(\Delta M)G/rc^2 = M' \qquad (9.3)$$

For $r \approx GM/c^2$, we have $M' = M$ (same as in GR). In the Newtonian case, $M' \Longrightarrow +\infty$, for $r \Longrightarrow 0$.

We can consider M and ΔM both negative (Case III). For a negative mass source G also changes sign (we shall also justify this in section 2 below) as we have universal repulsion. So:

$$M' = -M - \Delta M + G(\Delta M)M/rc^2 \qquad (9.4)$$

Also here as in (9.3), for $r \approx GM/c^2$, $M' = -M$. For $r \Longrightarrow 0$, the Newtonian case, $M' \Longrightarrow +\infty$.

Note that in Bonnor [9] the sign change of G for negative M was not taken into account leading to the erroneous conclusion that M′ becomes indefinitely negative whatever the value of r, a clearly absurd situation. So in case III general relativity would imply a maximum of $-Mc^2$, which is what one would expect from the mass symmetry of GR. We can also consider kinetic energy for test particles following trajectories in the above cases. For case I, of course we have $(\Delta M)v^2/2 - GM(\Delta M)/R$ for the total energy (in the non-relativistic approximation). So for $v^2=2GM/R$, we have a total energy of zero. Same result also holds for negatons test particles (i.e. $-\Delta Mv^2/2 + GM(\Delta M)/R = E_T = 0$ for $v^2 = 2GM/R$). So at a critical velocity (given by $v^2 = 2GM/R$) both positon and negaton test particles in the field of a positive mass have zero total energy. We can play around with the equations and get other interesting results.

Notice that we must distinguish negatons from tachyons to avoid confusion. Tachyons are particles with imaginary proper mass i.e. $m_o \Rightarrow im_o$ which always travel at a speed greater than light. Their minimum velocity is c and it would require an infinite amount of energy to reduce their velocity to c, i.e. their mass $\Rightarrow \infty$ as $v \Rightarrow c$ but from above. Their lowest energy (0) is at $v = \infty$ ($m = im_o/(v^2/c^2-1)^{1/2}$, $v \geq c$). Negatons however always have $v \leq c$, like usual positons but with simply m replaced by -m in relativistic dynamics, i.e. a real but negative proper mass. Tachions in contrast would have imaginary positive proper mass. In this paper we shall not consider tachyons.

Negatons interacting with positons must be accelerated because their energy must decrease. in this respect, negatons are similar to tachyons, higher velocity means more negative kinetic energy (hence lower energy) but of course v is restricted to be $v \leq c$, while for tachyons $v \geq c$. So if negaton energy decreases, that of positons must increase, by conservation of energy.

However the cross-section for interaction (for e.g. by

scattering) is very small. As indicated by Terletsky to avoid unbounded energy production we must have the constraint $n_+\sigma_{+-}R \sim 1$ since the energy gain of positons is of the form $E = E_o\exp(n_+\sigma_{+-}R)$ (usual cosmic ray type interactions). For neutron stars we have $n_+ \simeq 10^{38}cm^{-3}$, $R \simeq 10^6cm$, giving $\sigma_{+-} < 10^{-44}cm^2$, comparable with weak interaction cross-sections. However for larger σ_{+-} it is conceivable that this successive scattering of positons and negatons can lead to lage value of the exponent so $E >> E_o$ and perhaps the highest energy cosmic rays $\sim 10^{21}$ev can be produced. Negatons cannot be detected by standard methods because the absorption of a negative mass particle is equivalent to emission of a positive mass particle. This implies that a flux of negatons falling on a compact positive mass object can reduce its mass. We have the behaviour of some stars in the red supergiant stage which behave as if they have suddenly lost a lot of mass, usually attributed to very strong stellar winds. This accelerates their evolution (for e.g. Wolf-Rayet stars seem to have lost their envelope completely). There is some controversy in the usual models about stellar mass losses. Taking extreme case we can estimate the flux of negatons as

$$n_- \leq \rho\, R/mct \quad < \quad N/(\pi R^2 tc) \qquad (9.5)$$

R corresponding to the largest red giant radius $\approx 10^{14}cm$ (with their huge area they would capture more ambient negatons!), N $\approx$ total mass loss (number of particles) $\approx$ $M_\odot/m_p$ $\approx$ 10^{57}, t = evolution time. This gives $n_- < 1$ particle/cm^3 in interstellar space. A distribution of such negatons with this density would of course not be gravitationally bound into stars or galaxies (as they would mutually repel gravitationally and also repel positons!). Historically, Pearson [10] attributed the observed fast recession of a certain star to the fact that having negative mass it was being repelled from our region of space. These and other early speculations on negative mass are described in the book of Jammer [11].

The above estimated density of negatons would make them comparable in density to positons in interstellar space. In intergalactic space we would expect them to have same density of that of positons that is $\approx$ $10^{-6}cm^{-3}$. Such a uniform background density of positons and negatons can give rise to a long range gravitational screening analogous to Debye screening in electrostatics, in plasmas, where we have overall electrical neutrality, with equal numbers of positive and negative elctrical charges (force of a +ve electric charge being cancelled by a -ve charge). In the absence of negatons, (i.e. with only positive mass) such a gravitational shielding is impossible. In the Newtonian case the screening effect is described by the modified Poisson equation:

$$\nabla^2\Phi - \lambda_D^{-2}\Phi = -4\pi Gnm\delta(\bar{r}) \qquad (9.6)$$

Assuming a velocity of $\sim v_s \sim$ virial velocity in the galactic gravitational field $\sim 10^3$Km/sec:

$$\lambda_D \approx 2\pi v_s/(4\pi Gnm)^{1/2} \approx 1 \text{ kpc} \qquad (9.7)$$

this would give rise to a modification of the Newtonian gravitational potential (over Kpc distances) to

$$\Phi \approx 1/r - (1/r)\exp(-r/\lambda_D) \qquad (9.8)$$

In fact such a modification to Newtonian gravitational field over kiloparsec intergalactic distances has been proposed recently [12] by Sanders and others to account to resolve the dark matter problem and the problem of flat rotation curves of giant spiral galaxies. Again the presence of such a distribution of negatons over these distance scales, would prevent formation of galaxies possibly accounting for the large voids observed which are free of galactic or stellar systems. These may be some of the astrophysical consequences of negatons.

As noted early by Terletsky, negaton distributions must be thermodynamically associated with <u>negative</u> temperature as otherwise the partition function containing the Boltzmann distribution ($\sim \exp(-E/kT)$) would become singular (if $E < 0$, exponentially blow up unless $T < 0$ also). This is also consistent with the thermal energy of such particles also being negative, i.e. $-kT$.

However the entropy associated with the negatons would be positive. This comes about as $S = dE/dT \approx E/T$ and since E and T are both negative, S is positive. We shall see that this would assume significance when we discuss negative mass black holes and their quantum emission.

The action associated with these particles must also be negative as $I \sim \int mcds$ and for negative m, I is negative. This means that the quantum of action for negatons must be $-\hbar$ (i.e. negative Planck's constant). This is also supported by other reasoning. For instance the quantum uncertainty principle (which must also hold for negatons!) would imply a localisation length of $\sim \hbar/mc$. If m is negative, then clearly $\hbar$ must also be negative, since the Compton length must be positive. The energy transitions or quanta associated with negaton interactions would have energy of $E = \hbar c/\lambda$ and since λ is positive and E is negative, $\hbar$ should be negative.

Within this framework one could also picture the existence of negative energy photons with the associated Wien's law of blackbody radiation $\lambda_{max}T = \text{const} \approx hc/k$ since T would be negative for a black body radiation, $\hbar$ must be negative again. A negative temperature black body source would mean that as it emits negatons (or negative energy photons) its temperature would increase, i.e. it would become less negative. As we shall see soon, a negative mass black hole emitting thermal radiation behaves in just this manner. The entropy of the emitted black body radiation would however be positive, so that the second law of thermodynamics is not globally violated. Now the Schwarzschild metric is unaffected by a negative mass source (as both M and G reverse sign, GM remaining sign invariant). Thus

$$ds^2 = (1 - 2GM/rc^2)c^2dt^2 - dr^2/(1 - 2GM/rc^2) - r^2d\Omega^2 \quad (9.9)$$

still hold. So one could picture a negative mass black hole with a horizon ($r = 2GM/c^2 =$ positive). A positive mass black hole accreting negatons could presumably have its mass become negative. The effective mass of a Kerr-Newman black hole near the horizon is $M_{eff} = M - a^2/r - Q^2/r$; so M_{eff} can become negative if a^2, $Q^2$are sufficiently large. This also holds for Kerr hole with $M_{eff} = M - a^2/r$ or R-N hole with $M_{eff} = M - Q^2/r$.

Therefore black hole can become of effective negative mass.

However once we have a negative mass black hole it would not further capture any particles, all matter (both positons and negatons) being repelled by it. Its temperature would be negative, as $T_{bh} \simeq \hbar c^3/8\pi GkM$: GM is invariant, but $\hbar$ is negative so T is negative. However the entropy of a negative mass black hole would still be positive, as $S \approx (GM^2/\hbar c)k$. GM is invariant but M is negative and $\hbar$ is negative; so S is positive. This implies that a negative mass black hole cannot emit positive energy particles If they emit positive mass particles, their mass would become more negative.

Moreover a black body of negative temperature T would emit a negative energy flux in accordance with Stefan's law ($E = aT^4$, T^4 positive but $a = 8\pi^2k^4/45h^3c^2$ is negative!). So we have consistency of argument. As entropy of emitted negative energy particles is positive, as argued before, the entropy of the hole decreases, like the usual positive mass hole, its area reduces and it evaporates. Its mass becomes less negative, i.e. more positive. It may tend to a stable positive Planck mass $M_{Pl} \simeq (\hbar c/G)^{1/2}$ [13]. Note that M_{Pl} has to be positive as both $\hbar$ and G are negative! So both black holes of positive and negative mass may tend to have a stable remnant of positive mass around M_{Pl}.

The lifetimes of both positive and negative mass black holes with the same mass are equal ($t = G^2M^3/\hbar c^4$, t is positive for both cases!), but there is a possible way to distinguish the sign of the black hole mass: in fact a negative mass black hole does not gather matter but emits Hawking radiation of negative energy particles. The angular momentum associated with a negative mass would also be negative ($J = MvR$, M negative, so J is negative). The

Schuster's law would imply $J \approx (G/v)M^2$ relation. As G reverses sign for negative M, J is negative (M^2 is positive). The magnetic moment given by $m = \beta J/c$ would also be negative for negatons.

As also remarked by Bonnor, a Bohr atom of negatons is possible only for electric charges of the same sign. As $-|m|v^2/R$ = force between charges is negative (that is repulsive force), the potential must be positive. For unlike charges we have attractive force so potential is negative and force is positive. (Note that negatons move opposite to the direction of the force).

However the electrical properties of negatons are bound to be peculiar (and Bonnor argument may not hold): if we admit a classical radius, e^2 must be negative (classical radius $r_e \sim e^2/mc^2$) so charges must be $\pm ie$, i.e. the electromagnetic phase $\exp(ie\phi)$, must be orthogonal with respect to positons (for instance if we have a current of charged negatons and external magnetic field is applied, current will be deflected in an opposite manner to the usual way [Lenz-law] and also polarisation of the emitted radiation would have different properties). This also implies that Larmor (dipole) radiation from a charged negaton which is accelerated would carry negative energy, as $E_{em} \sim 2e^2a^2/3c^3$, (with e^2 negative).

This is also true for gravitational quadrupolar radiation $E \sim GM^2R^4\omega^6/c^5$. As M^2 is positive, G is negative, E must be negative. So all radiation emitted by a negaton, whether electromagnetic or gravitational, must carry negative energy. This implies that the negaton mass does not indefinitely increase and catastrophically release an infinite amount of energy as would happen if it emits radiation carrying positive energy. In this case we should be violating thermodynamics (getting indefinite energy from a finite amount).

For this reason, the result of Hoffmann must be considered unphysical if not erroneous. With our consistent treatment we do not meet with this difficulty. All this is also supported by another result. The dipole power radiated by a sphere with radius r oscillating back and forth with amplitude A and frequency ω in a medium of density ρ, is $\langle\omega_{rad}\rangle_{av} = 2\pi\rho\omega^6 A^2 r^6/27c^3$, which for a negative density medium is clearly negative.

The uncertainty priciple would of course permit the multiproduction of particles of both positive and negative mass out of the vacuum. The way to search for negative mass particles in high energy collisions is to look for calorimetric details of missing energy. In fact in many event with jet production in high energy collisions, energy anomalies such as in transverse momenta have been noted. Missing energy need not necessairly be carried away from unknown particles such as these postulated in supersymmetry but could be also due to production of negatons. However the cross-sections are expected to be very low.

In a universe with equal amounts of negatons and positons, the net mass is zero. In fact a R-W universe with k = 0, i.e. spatially flat, has zero total energy as is well known. This could be a possible model in such a case.

2. Dirac equation with negative mass

Relativistic wave equations for particles with any spin are generally form-invariant for particles of both positive and negative mass. This is true for the Klein-Gordon equation, Proca equation, linearized massive spin-2 Pauli-Fierz equation, the Dirac equation etc. Here we shall be concerned chiefly with the Dirac equation. The Dirac equation is well known to contain negative energy solutions. In fact Dirac predicted energy states being occupied, with a hole, in the sea, corresponding to a positron. In quantum mechanics it is known that the centre of a wave packet of negative energy solutions describes a uniform motion with velocity in opposite direction to momentum thus behaving like a negative mass particle [14]

As far as spin-1/2 particles are concerned it is the chirality transformation described by $\psi \Rightarrow \gamma_5\psi$, which is the similarity transformation, which converts the Dirac equation for positive mass, i.e. $(\gamma^\mu\nabla_\mu + m)\psi = 0$ into the one for negative mass, i.e. $(\gamma^\mu\nabla_\mu - m)\psi = 0$, i.e. if $\psi(x)$ is a solution for the positive mass case, then $\gamma_5\psi(x)$ is a solution of the negative mass Dirac equation [15].Here γ_5 has the usual properties: $\gamma_5^2 = 1$, $\nabla_\mu\gamma_5 = 0$. Thus the chirality transformation has the effect equivalent to mass inversion: $m \Rightarrow - m$ (so that this transformation either changes sign of the charge or of the mass), and moreover as is well known $\psi \Rightarrow \gamma_5\psi$, interchanges helicity, i.e. helicity is flipped ($h \Rightarrow - h$). This shows that parity inversion and mass reversal are equivalent. Under the chirality transformation, $\bar{\psi}\psi$ (which represents the density) changes sign; i.e. it converts $T_{\mu\nu}(x) \Rightarrow - T_{\mu\nu}(x)$. The formalism is carried over into the curved-space Dirac equation, where we have the integrability condition (∇_μ are covariant derivative):

$$(\nabla_\nu\nabla_\mu - \nabla_\mu\nabla_\nu)\psi = (1/4)\, R_{\mu\nu\sigma\rho}\gamma^\rho\gamma^\sigma\psi \qquad (9.10)$$

with

$$\nabla_\mu\psi = -(1/4)\, m\, \gamma_\mu\psi \qquad (9.11)$$

and

$$R_{\mu\nu\rho\sigma} = K(g_{\mu\rho}g_{\nu\sigma} - g_{\mu\sigma}g_{\nu\rho}) \qquad (9.12)$$

$K = (1/3)\,\chi\,\rho$ (for spaces of maximal symmetry).

Manipulation of the above equations gives:

$$\chi\,\rho = (3/4)\, m^2 \qquad (9.13)$$

As m^2 is positive, it follows that if ρ is transformed as $\rho \Longrightarrow -\rho$ the sign of χ and hence G must be reversed. This is consistent with the results in section 1.

One can obtain a spectrum of negative masses, in the same way as positive masses, by considering the form invariance of the Dirac equation in flat spaceunder <u>discrete</u> <u>scale</u> transformations. Now the Lagrangian

$$\mathcal{L} - \bar{\psi}\,\gamma^{\mu}\partial_{\mu}\psi \quad - \quad m\,\bar{\psi}\,\psi \tag{9.14}$$

has dimension $l = -3/2$ in absence of interactions under scale transformations:

$$x_{\mu} \Rightarrow x'_{\mu} \equiv \varepsilon\, x_{\mu}\,, \qquad \psi(x) \Rightarrow \psi'(x') = \varepsilon^{-3/2}\psi(\varepsilon x) \tag{9.15}$$

under continuous scale transformations, i.e. with continuous ε. Now consider discrete scale transformations which in general are:

$$x_{\mu} \Rightarrow x'_{\mu} \equiv \beta^{n}x_{\mu} \quad (n = \ldots.,2,1,0,-1,-2,\ldots\ldots) \tag{9.16}$$

then

$$\psi(x) \Rightarrow \psi'(x') = \beta^{-3n/2}\psi(\beta^{n}x) \tag{9.17}$$

Here β is not variable but takes some fixed value (any number). Then $\mathcal{L}$ transforms as

$$\mathcal{L} \Rightarrow \mathcal{L}' = [- \psi' \gamma^{\mu}\partial'_{\mu}\psi' - (m \beta^{n})\bar{\psi}'\psi] \qquad (9.18)$$

i.e. same form but with m replaced by m β.

Thus this means that the existence of a Dirac field with quanta of mass m implies the existence of field with quanta $m_n = m_o\beta^n$, where n can be a positive or a negative integer. So for odd value of n with negative β we would have <u>negative</u> masses with $m_n = m_o\beta^n$

3. <u>Hamiltonian formalism</u>

It is known that the general structure of metric-torsion theories allows parity violating interactions implying that the torsion of space-time might be responsible for parity-violation in weak interaction. It has been recognized (see Chp.V and ref.[16], [17]) that the torsionic contact interaction between two Dirac particles has a formal analogy with the weak interaction Lagrangian in the usual Einstein-Cartan static term for the field, i.e. the Dirac equation can be written in this framework as [18]:

$$i\gamma^{o}\partial_{o}\psi = - i\gamma\cdot\nabla\psi + m \psi - (3\chi/8) \bar{\psi} \gamma_{n}\gamma_{5}\psi \gamma_{n}\gamma_{5}\psi \qquad (9.19)$$

In the Hamiltonian formulation, we may write

$$H = H_{free} + H_{int} \qquad (9.20)$$

where

$$H_o = \int_V dV \; [\tfrac{1}{2} \, (-i \, \bar{\psi} \, \gamma \cdot \nabla\psi + h.c.) \; + \; m \, \bar{\psi} \, \psi] \qquad (9.21)$$

(V = volume of region)

$$H_I = -(3\chi/32)\int_V dV \; \bar{\psi} \; (\bar{\psi} \; \gamma_n\gamma_5\psi)\gamma_n\gamma_5\psi \qquad (9.22)$$

(i.e. same form as the neutral current interaction [19]) with the equal time commutation relation:

$$\left\{ \psi \, (r,t), \; \bar{\psi} \, (r',t) \right\} = \gamma_o \; \delta \, (r - r') \qquad (9.23)$$

The eigenstate of H can be formulated in the usual way in the second quantization formalism; i.e. if $u_\sigma(\hat{k})$, $v_\sigma(\hat{k})$ are spinors with positive and negative energy and $a_{\sigma\hat{k}}$, $b^{*}_{-\sigma,-\hat{k}}$ are annihilation operator for particles of momentum k (helicity σ) and creation operator for antiparticles respectively, then

$$\psi = \frac{1}{V^{1/2}} \sum_{\sigma,\hat{k}} \{a_{\sigma\hat{k}} \; u_\sigma(\hat{k}) \; + \; b^{*}_{-\sigma,-\hat{k}} \; v_\sigma(\hat{k}) \; \} \qquad (9.24)$$

In terms of the occupation numbers for particles and antiparticles, i.e. $n^{(+)}_{\sigma,\hat{k}}$, $n^{(-)}_{\sigma,\hat{k}}$, the energy eigenvalue of H_o for the state is:

$$|\phi\rangle = |\{ n^{(+)}_{\sigma,\hat{k}}, n^{(-)}_{\sigma,\hat{k}} \}\rangle \qquad (9.25)$$

$$E_{\circ} = \sum_{\sigma\hat{k}} (k^2 + m^2)^{1/2} (n^{(+)}_{\sigma\hat{k}} + n^{(-)}_{\sigma\hat{k}}) \qquad (9.26)$$

The vacuum expectation value of ϕ is given by [20]:

$$\langle\phi|\bar{\psi}\ \gamma_n\gamma_5(\bar{\psi}\ \gamma_n\gamma_5\psi)\psi|\phi\rangle = (\mathrm{Tr}\gamma_n\gamma_5 A)(\mathrm{Tr}\gamma_n\gamma_5 A) - \mathrm{Tr}\gamma_n\gamma_5 A\gamma_n\gamma_5 A \qquad (9.27)$$

where

$$A = \frac{1}{V}\sum_{\sigma\hat{k}}\{u_\sigma(\hat{k})\bar{u}_\sigma(\hat{k})n^{(+)}_{\sigma\hat{k}} - v_\sigma(\hat{k})\bar{v}_\sigma(\hat{k})n^{(-)}_{\sigma\hat{k}}\} \qquad (9.28)$$

We consider an asymmetric state with respect to C - operator, i.e. different coupling to particles and antiparticles thus (for instance if torsion couples to isospin [21]):

$$n^{(+)}_{\sigma\hat{k}} = n^{(+)}(|\hat{k}|) , \quad n^{(-)}_{\sigma\hat{k}} = n^{(-)}(|\hat{k}|) \qquad (9.29)$$

giving asymmetry in occupation number:

$$\sum (n^{(+)}_{\sigma\hat{k}} - n^{(-)}_{\sigma\hat{k}}) = N^{(+)}_B - N^{(-)}_B \neq 0 \qquad (9.30)$$

Here $N^{(+)}_B$ = total baryon number

$N^{(-)}_B$ = total antibaryon number.

The number densities of baryons and antibaryons in this case are

$$n^{(+)} = \frac{1}{V}\sum_{\sigma\hat{k}} n^{(+)}_{\sigma\hat{k}}\ , \quad n^{(-)} = \frac{1}{V}\sum_{\sigma\hat{k}} n^{(-)}_{\sigma\hat{k}}\ , \quad n = n^{(+)} - n^{(-)} \qquad (9.31)$$

$$A = (\gamma_o/4)(n^{(+)} - n^{(-)}) + (m/(k^2 + m^2)^{1/2})\, n/4 \qquad (9.32)$$

The vacuum expectation value of the Hamiltonian density is then:

$$\frac{1}{V}\langle\phi|H|\phi\rangle = E_o/V - (3\chi/32)(n^{(+)} - n^{(-)})^2 \qquad (9.33)$$

We had argued [22], that at a sufficiently high density, spin effects become dominant in the early universe for the Einstein-Cartan cosmology wherein the effective density can become negative i.e. :

$$\rho_{eff} = (\ \rho - (2/3)(\pi G\sigma^2/c^4)) \qquad (9.34)$$

The above expression for the Hamiltonian density also suggests that at sufficiently high density (we know in fact that $n^{(+)} >> n^{(-)}$ in the universe) the energy density can become negative, giving an effective negative mass and consequently an inflationary expansion. Conversely the inflationary expansion induced by torsion [23], can also give rise to a matter-antimatter asymmetry (because of different coupling of torsion to isospin [21]). In this case this would be the only scenario, where inflation and matter-antimatter asymmetry are both produced by a single mechanism, i.e. that involving spin-isospin effect due to torsion. This can also lead to segregation of particles and antiparticles. We can also consider an asymmetric state with respect to parity; in this case the H density becomes:

$$\frac{1}{V} <\phi|H|\phi> = E_o/V - (3\chi/8)n^2 \qquad (9.35)$$

and for sufficiently large number density n thisimplies a negative energy density.

For the Robertson-Walker metric we have the usual equations:

$$3kc^2/a^2 + 3\dot{a}^2/a^2 = \chi <T^o_o> \qquad (9.36)$$

$$kc^2/a^2 + \dot{a}^2/a^2 + 2\dot{a}^{\cdot}/a = \chi <T^1_1> \qquad (9.37)$$

and

$$<T^1_1> = <T^2_2> = <T^3_3>$$

For the above interaction of spins:

$$<T^o_o> + 3<T^1_1> = m<\bar{\psi}\psi> + (3\chi/8) <\bar{\psi}\gamma_n\gamma_5\psi><\bar{\psi}\gamma_n\gamma_5\psi> \qquad (9.38)$$

If N is total number of particles:

$$<\bar{\psi}\gamma_n\gamma_5\psi><\bar{\psi}\gamma_n\gamma_5\psi> \approx N^2/a^6 \qquad (9.39)$$

Thus:

$$\langle T^o_o\rangle = (\mathrm{const}/a^4) - (3\chi/16)\ \langle\bar{\psi}\gamma_n\gamma_5\psi\rangle\langle\bar{\psi}\gamma_n\gamma_5\psi\rangle + (m/2)\langle\bar{\psi}\psi\rangle \tag{9.40}$$

Thus $\langle T^o_o\rangle$ can be effectively negative at sufficiently high N and sufficiently small 'a' (~ N^2/a^6). The strong dependence on 'a' of the torsion term is to be noticed. In general:

$$(1/V)\langle H\rangle = (1/V)E_o - (3\chi/16)\ \langle\bar{\psi}\gamma_n\gamma_5\psi\rangle\langle\bar{\psi}\gamma_n\gamma_5\psi\rangle \tag{9.41}$$

the second term going as N^2/a^6, becoming large at small 'a'. This would give a minimum value [24] for the radius of a collapsing universe, suggesting the possibility of a bounce, due to the effective density becoming negative. This would also happen in the case of a black hole giving a stable remnant supported by torsion [13].

We notice torsion effects in supergravity involving higher, spin 3/2 fields, like that of gravitino also involve torsion terms analogous to the above. For instance, the spin-3/2 contorsion tensor given by:

$$K_{\mu ab}(\psi_a) = (\chi^2/4)\ (\bar{\psi}_\mu\gamma_a\psi_b - \bar{\psi}_\mu\gamma_b\psi_a + \bar{\psi}_a\gamma_\mu\gamma_b) \tag{9.42}$$

gives rise to an extra spin-interaction term in the Hamiltonian o the type

$$(-e\chi^2/32)\left[(\bar{\psi}^{\mu}\gamma^{\nu}\psi^{\rho})(\bar{\psi}_{\mu}\gamma_{\nu}\psi_{\rho} + 2\bar{\psi}_{\nu}\gamma_{\mu}\psi_{\rho}) - 4(\bar{\psi}_{\mu}\gamma_{\nu}\psi_{\rho})^2 \right] \qquad (9.43)$$

(where e is gauge coupling).

Such term would again gives rise to a negative mass interaction in the early universe. In short, <u>torsion</u> provides a <u>natural</u> framework for the description the <u>negative</u> mass under extreme conditions of such as in the early universe, when a transition from positive to negative mass can take place.

<u>Reference</u>

[1] Ya.P.Terletsky - Journ.Phys.Rad.<u>23</u>, 91, (1962); Ann.Inst.Poincaré <u>1</u>, 421, 431 (1964); reprint of Erice Intern.School of Cosmol.and Grav. 1975; see also "Paradoxes in the theory of Relativity" Plenum Press, New York, 1968

[2] H.Bondi - Rev.Mod.Phys.<u>29</u>, 3, 423 (1957); J.Weber in "General Relativity and Gravitational Waves" Interscience Publisher, New York, 1961 p.5; M.O.Costa de Beauregard - C.R.<u>249</u>, 1737 (1961); A.Inomata and D.Peak - Il Nuovo Cimento <u>63B</u>,132 (1969)

[3] P.Morisson - Amer.Journ.Phys. <u>26</u>, 358 (1958); see also J.W.Gardner - New Scientists <u>42</u>, 578 (1969)

[4] T.Goldman and M.M.Nieto - Phys.Letters <u>112B</u>, 437 (1982); N.Beverini et al. - preprint CERN/PSCC/86-2

[5] L.I.Schiff - Phys.Rev. Letters, <u>1</u>, 7, 254 (1958)

[6] E.A.Uehling - Phys.Rev. <u>48</u>, 55 (1935); R.Serber - Phys.Rev.<u>48</u>, 49 (1935)

[7] F.Winterberg - Il Nuovo Cimento, <u>19</u>, 186 (1961)

[8] V.de Sabbata and C.Sivaram - in "Theoretical and Experimental Problems of Gravitation" pag.89 (1992)

[9] W.B.Bonnor - Gen.Rel.and Grav. 21, 1143 (1981)

[10] K.Pearson - Ann.Journ.Math. 13, 309 (1981)

[11] Max Jammer - "Concept of Mass", Harward College 1961 (italian translation, Feltrinelli, Milano 1974)

[12] R.H.Sanders - MNRAS 223, 539 (1986); Astron.Astrophys. Rev.2, 1 (1990)

[13] V.de Sabbata and C.Sivaram - "The central role of spin in black hole evaporation" in 'Black hole physics' Kluwer Academic Publ. NATO ASI Series, Vol.364, 225 (1992)

[14] A.Messiah - "Quantum Mechanics", Vol.2° Amsterdam 1962 p.952

[15] J.J.Sakurai - Il Nuovo Cimento, 7, 649 (1958)

[16] V.de Sabbata and M.Gasperini - Lett.Nuovo Cimento 21, 328 (1978)

[17] V.de Sabbata and M.Gasperini - Gen.Rel.Grav.10, 731 (1979)

[18] V.de Sabbata and M.Gasperini - "Introduction to Gravitation" World Sci. 1985 p.249

[19] E.Lifchitz et L.Pitayevski - Théorie quantique rélativiste" Vol.II ed.Mir de Moscou 1973

[20] Schweber - "An introduction to relativistic quantum field theory", Harper & Row New York 1964

[21] V.de Sabbata and C.Sivaram - Intern.Journ.of Theoret. Phys.29, 1 (1990)

[22] V.de Sabbata and C.Sivaram - Astrophys.Spac.Sci.165,51 (1990)

[23] V.de Sabbata and C.Sivaram - Astrophys.Spac.Sci.176, 141 (1991)

[24] A.Trautman - Nature 242, 7 (1973)

CHAPTER X

SYMMETRY BREAKING

Torsion as length defects (dislocations); Curvature as angle defects? (disclinations?)

1. Torsion and symmetry breaking

In this section we shall consider the gauge invariance and spontaneous symmetry breaking associated with the above formalism. The invariance under tetrad rotations has already been noted. This corresponds to invariance under the gauge group SL(2,C) and is analogous to invariance of the Yang-Mills field under local isotopic-spin rotations, i.e. under SU(2). As noted already the transformation laws for the fields γ^{μ} and Γ_{μ} under group of tetrad rotations (i.e. under SL(2,C) denoted by S):

$$\gamma^{\mu} \Longrightarrow S\gamma^{\mu}S^{-1},\ \Gamma_{\mu} \Longrightarrow S\Gamma_{\mu}S^{-1} + (\partial_{\mu}S)S^{-1},\ \Gamma_{\mu\nu} \Longrightarrow S\Gamma_{\mu\nu}S^{-1} \qquad (10.1)$$

where $\Gamma_{\mu\nu} = \partial_{\nu}\Gamma_{\mu} - \partial_{\mu}\Gamma_{\nu} + [\Gamma_{\nu},\Gamma_{\mu}]$. If T denotes the space-time dependent isotopic rotation, we have for the Yang-Mills field B_{μ} interacting with fields having non-vanishing isotopic spin, the transformation laws:

$$B'_{\mu} = T^{-1}B_{\mu}T + (\partial_{\mu}T)T^{-1},\ F'_{\mu\nu} = T^{-1}F_{\mu\nu}T,\ \psi' = T\psi T^{-1} \qquad (10.2)$$

where $F_{\mu\nu} = \partial_{\nu}B_{\mu} - \partial_{\mu}B_{\nu} + [\Gamma_{\nu},\Gamma_{\mu}]$. The analogy clearly indicates that the field $\Gamma_{\mu\nu}$ transforms like a Yang-Mills field under the gauge group SL(2,C). As far as gravity is concerned, the point to note is that there is no distinction between its internal symmetry group and the group of general coordinate transformations in space-time. As stated before there is no difference between the reference frame and its space-time dependence. The transformation matrix for the SL(2,C) group can be chosen up to a phase as:

$$S = \exp\left[(1/4)\chi\, \theta_{ab}\gamma^a\gamma^b\right] \qquad (10.3)$$

Γ_μ are then the gauge potential and $\Gamma_{\mu\nu}$ are the gauge fields. All the equations in above section reflect SL(2,C) invariance. For instance the Lagrangian for the generally covariant Dirac equation (see eq.5.23)

$$\mathcal{L} = (i/2)t[\bar{\psi}\, \gamma^\mu(\partial_\mu + i\Gamma_\mu)\psi + \Lambda^{1/2}\bar{\psi}\, \psi] + h.c. \qquad (10.4)$$

(with the SL(2,C) covariant Γ_μ) is SL(2,C) invariant. The Einstein-Cartan equations may be expressed in the more elegant forms [1]:

$$i[\gamma^\mu, \Gamma_{\mu\nu}] = \chi J_\mu \;;\quad \nabla_\mu\, [\gamma^\mu, \gamma^\nu] = \chi S^\nu \qquad (10.5)$$

in terms of a conserved set of SL(2,C) Noether currents: $J^\nu = S^\nu - i[\Gamma_\mu, [\gamma^\mu, \gamma^\nu]]$, which are identically conserved:

$$\partial_\mu J^\nu = \partial_\nu \partial_\mu [\gamma^\mu, \gamma^\nu] = 0 \qquad (10.6)$$

<u>Torsion is necessary</u> for this invariance. The Higgs-Kibble

mechanism is naturally present in the SL(2,C) gauge invariant Einstein-Cartan-Weyl-Dirac Lagrangian. The Higgs machanism is an exception to the Goldstone boson theorem (with states that whenever we have spontaneous breaking of a symmetry we must have accompanying <u>massless</u> fields and the maximum number of such fields will correspond to the number of broken generators connected with the broken symmetry) when the Lagrangian exhibits local gauge invariance such as in the above case. One can linearize the Lagrangian by writing: $\gamma^{\mu a} = \eta^{\mu a} + \phi^{\mu a}(x)$ and discarding quadratic terms above. The ground state solution is then given by: $<\gamma^{\mu}> = \gamma_a \eta^{\mu a}$; which is <u>not</u> SL(2,C) invariant, i.e. we have a sixteen-component field γ^{μ} with a non-zero vacuum expectation value (VEV). As a result of this spontaneous symmetry breaking, we would have six Goldstone boson modes corresponding to the six broken SL(2,C) generators. Thus we do not need a Higgs field to break the symmetry and give the particle masses. The situation here is analogous to that of a ferromagnet when the rotational invariance O(3) is broken below the Curie point when the spins become aligned and a net magnetisation results. Above the Curie temperature, the symmetry is restored . The analogy becomes stronger if we consider that the SL(2,C) invariance of the torsion Lagrangian is isomorphic to O(3,1) invariance which is again a rotation group in four dimensions.

So for understanding the origin of neutron star magnetic field, this seems to indicate that above temperatures $T_h \approx$

10^{13} K, the neutron star magnetic field may vanish or diminish considerably. Such heating may be brought about by rapid accretion on an already existing neutron star or collapse of white dwarf. Millisecond pulsars which are known to have rather low magnetic fields might have undergone such a transition which makes them spin much faster but also removes magnetic field. The analogy with the Curie temperature would also indicate a susceptibility which is inversely proportional to temperature. Indeed the torsion induced magnetic moments of elementary particles show a temperature dependence $\propto T^{-1}$ [2]. This is also brought about by G scaling with temperature as T^{-2} in the early universe. Thus phase transitions in the early universe would also be accompanied by generation of a primordial magnetic field as the universe cools below the phase transition temperature when the spin-torsion interaction becomes effective. The question of symmetry breaking and mass generation also brings in the analogy with type II superconductors. Here phase invariance is broken (i.e. no longer invariance under $\exp(i\phi)$)in the superconducting ground state, the photon acquires an effective mass (related to the penetration depth) The symmetry is restored above a critical value of the magnetic field strength. The critical value of the magnetic field which restores the symmetry is given by

$$B_c \simeq m^2c^3/2e\hbar \qquad (10.7)$$

(2e is the charge carried by the Cooper pairs, m is the mass induced). For a typical superconductor, with a penetration depth of a few microns, the critical field is a few Tesla (corresponding to an effective mass $\approx$ 10^{-3}ev). As torsion induces a magnetic field when the invariance under SL(2,C) i.e. $\exp\{(1/4)\chi\theta_{ab}\gamma^a\gamma^b\}$ is broken, the corresponding effective mass induced would be obtained by equating the magnetic fields in eqs.(10.7) and (2.27), i.e.:

$$(m^2c^3/e\hbar) = (8\pi/3c)(2\alpha G)^{1/2}\sigma \qquad (10.8)$$

As the weak interaction is already contained in the generally covariant Dirac equation as noted in section 2 of Chp.V, we can apply eq.(10.8) to calculate the induced mass of the weak intermediate boson. For this, G is replaced by G_F, the Fermi constant, and in estimating the spin density, the volume is considered to be the beta decay length cubed. Thus we have: (see also (8.12))

$$(m_w^2c^3/e\hbar) = (8\pi/3c)(2\alpha G_F)^{1/2}\{\hbar/(G_F/\hbar c)^{3/2}(4\pi/3)\}(c/\hbar) \qquad (10.9)$$

We note that α and G_F occur in the same formula suggesting electroweak unification. Eq.(10.9) gives:

$$m_w \approx 10^{-22}g \approx 70 \text{ Gev} \qquad (10.10)$$

which agrees well with the observed mass of the $W^{\pm}$ boson. Here we had not used a Higgs scalar field to generate mass. The analogy with superconductivity as used to generate an effective mass below the critical electroweak magnetic field induced by torsion. At very high temperatures and large magnetic fields above the critical field (B scales as T^2 for instance in the early universe)the effective mass vanishes and the W boson becomes massless, symmetry is restored and electroweak unification occurs. So high temperature and high magnetic field are equivalent, symmetry is restored above the critical temperature and/or magnetic field. It is broken below the critical magnetic field and/or temperature. The above discussion is also relevant at other energy scales. When applied to strong interactions, we must use the appropriate strong gravity constant G_f in eq.(10.9). When this is done, we get the induced mass as $m_f \approx 1$Gev, the mass of the f-meson. So strong interactions arise when symmetry is broken around 1Gev (temperature has the critical value $\approx 10^{13}$ °K as noted above and the corresponding critical magnetic field is $\approx 10^{19}$Gauss). For the Newtonian constant G_N, eq.(10.9) would give the induced mass due to symmetry breaking as that of the Planck mass M_{Pl} as is to be expected! The coresponding temperature and field would be ~ 10^{32} °K and 10^{57}G as already noted. Eq.(10.9) would once again suggest the universal scaling relation [3,4] which we have already noted several times, i.e.(see for instance eq.(8.11)):

$$G_f m_f^2 = G_N m_{Pl}^2 = G_W m_w^2 \qquad \text{etc.} \qquad (10.11)$$

So one can now understand how symmetry breaking at

different scales can generate coupling constants and masses corresponding to the different fundamental interactions all from a single underlying framework. Moreover it follows that we can easily generalize the SL(2,C) gauge invariant Einstein-Cartan-Weyl-Dirac Lagrangian to include internal symmetries such as SU(3), SU(6) etc. [1].

To describe for instance the nonet of strongly interacting spin-2 mesons which involves the hadronic symmetry group SU(3) the simplest choice of the gauge group would be SL(2,C) **x** SL(6,C). By analogy with the fields, γ^{μ} and Γ_{μ} of the SL(2,C) case, we would have the fields M^{μ} and H_{μ} given by:

$$
\begin{aligned}
M^{\mu} &= M^{\mu a i}(\gamma_a \lambda_i/2) + M^{a i 5}(i\gamma_a \gamma_5 \lambda_i/2) \\
H_{\mu} &= H_{\mu}^{a b i}(\gamma_a \gamma_b \lambda_i/4) + H_{\mu}^{i}(\lambda_i/2) + H_{\mu}^{5 i}(\lambda_i \gamma_5)
\end{aligned}
\tag{10.12}
$$

λ_i are the Gell-Mann matrices of SU(3), i.e. the fields are generalized to contain (4**x**72) components each rather than the (4**x**4) components of SL(2,C) case; we now deal with "72-beins" rather than vierbeins! Then the Einstein-Cartan -Weyl-Dirac-GellMann Lagrangian is simply given by replacing the γ^{μ} and Γ_{μ} for the SL(2,C) case by the M^{μ} and H_{μ} for SL(6,C) invariant case, i.e. the generally covariant Dirac equation has the same form as (5.23) but with γ^{μ} and Γ_{μ} replaced by M^{μ} and H_{μ}. For the very general case SL(2n,C) case, the set of gauge potentials H_{μ} are given as :

$$H_\mu = \sum_{j=1}^{N}(H^j_\mu + iH^{5j}_\mu\gamma_5)(\lambda_j/2) + \sum_{j=1}^{N}(1/4)H^j_{\mu ab}\gamma^a\gamma^b\lambda_j \qquad (10.13)$$

These would replace Γ_μ in the formalism of section 2 of Chp.V i.e. the generalized Dirac equation (5.23) which is SL(2,C) invariant can be made SL(2,C)**x** SL(2n,C) invariant simply by replacing the covariant derivative $(\partial_\mu + i\Gamma_\mu)$ in (5.23) by a derivative $(\partial_\mu + i\Gamma_\mu + iH_\mu)$, H_μ given as in (10.13). For n = 3, one would get the SL(2,C)**x** SL(6,C) invariant Dirac equation which is sufficient to describe the strong gravitational interaction of the nonet of spin-2 mesons with the basic hadronic fermionic fields, i.e. one would now use the H_μ given in eq.(10.12). The extra terms in these covariant derivatives will now generate a whole variety of interaction terms involving hadronic currents and their products, consisting of vector, axial vector, pseudoscalar and other terms analogous to the SL(2,C) case. One can thus describe hadronic torsion in a manner analogous to that explained in sections 2 and 3 of Chp.V. We can again use the symmetry breaking mechanism for this general case also to generate the various particle masses.

2. Curvature from spin

We have seen before (section 4 of Chp.IV; see also Chp.XI) that the dislocation defect induced by torsion in space time [5], [6] due to the presence of intrinsic spin behaves like a string with a tension, i.e.mass per unit-length determined by the torsion Q as $\int QdA$ and this can be quantized (for existence of quantized values of the spin) through the Einstein-Cartan equations in integral multiples of the Planck length etc. The defect length when multiplied by the tension then gives the mass. Now as is known from the geometry of strings, the string tension can be related to the defect angle as

$$\Delta\theta \approx 2\pi\ (1 - 4G\mu/c^2) \qquad (10.14)$$

μ being the tension. The deviation from Minkowski space is locally given by the defect angle $\Delta\theta$ (i.e. $0 \le \theta \le 2\pi(1 - 4G\mu/c^2)$. For small μ the deviation becomes smaller and smaller, i.e. the curvature is small.

Now the defect angle is related to the mean curvature through the relation

$$\Delta\theta = \int R\ dA \qquad (10.15)$$

which implies for positive curvature that we have an excess over 2π for a triangle drawn on the surface. $\int R\ dA$ gives the integrated curvature over the surface. Thus we have the sequence:

spin $\Longrightarrow$ torsion $\Longrightarrow$ defects in space-time $\Longrightarrow$ symmetry-breaking $\Longrightarrow$ tension $\Longrightarrow$ mass $\Longrightarrow$ deficit angle $\Longrightarrow$ curvature (of course this is a further contribution to the curvature which may already exist).

Also it has been pointed out in the recent work of Fujiwara et al. [7] that masses are proportional to deficit angles around particle trajectories. They give an important relation between mass and deficit angle as:

$$\mu = c^2\Delta/8\pi G \qquad (10.16)$$

From the definition of Gaussian curvature in terms of mass density as $R = 8\pi G\rho/3c^2$, this is equivalent to (10.15), the integration of ρ giving mass. Also Fujiwara et al. point out that pasrticles are created in negative energy regions as defects in space-time, or equivalently in a space time with negative cosmological constant.

As shown in Chp.IX (see also [8],[9],[10],[11]) the expectation value of the Hamiltonian for spinor fields interacting with torsion can become negative, and this implies that torsion also leads to regions [11] with negative energy density without any ad hoc assumption (or equivalently to a negative cosmological constant). We should not be surprised at the similar result with torsion; i.e. to formation of defects with a deficit angle related to tension and hence mass and curvature which would lead to creation of particles in such regions. Thus we have consistency with the work of Fujiwara et al. who start from a different premise.

Eq.(10.15) can also be interpreted as the phase acquired,

i.e. a kind of geometric phase acquired, by an amount $\Delta\theta$, when a test particle moves around a distribution of dislocations defects in space-time, induced by torsion. This is equivalent to moving in an enclosed region having an average curvature R. If R is the curvature varying from point to point, it is the integrated curvature over the area element $\int R\ dA$ which contributes to the phase, i.e. the deviation from 2π when the particle makes a single path. If we have a closed path, for n turns the phase acquired should be n $\Delta\theta$. Now, as dislocation defects occur in crystal, one can think of an experimentally testable situation, when particles such as neutrons on moving through a distribution of such defects acquire a phase; this phase can be possibly measured and would then give information on the topological property of the deformed space around these defects as measured by $\int R\ dA$ (see [12]). In other words some property like curvature can be associated with the space around the distribution of defects. Kawamura [13] considered 'phase dismatching' in dislocated crystals.

Thus presence of torsion in space caused by spin, must necessarily also lead to curvature. In other words a space with torsion and curvature can be considered completely equivalent to an elastic continuum which has undergone plastic deformations being full of dislocations and disclinations, which will play a role analogous to a string. In the case of a cosmic string it is known that a quantum particle moving around the string acquires an Aharonov-Bohm phase caused by the change in the global topology of space-time induced by the presence of the string. Such defects and the corresponding effects associated with, then should in principle also exist in planar elastic materials. Diffraction of high energy particles from screw dislocations should in principle show such effects.

Sakharov considered space-time to be analogous to an elastic continuum, with the *gravitational* constant defining the metric elasticity. However he did not consider plastic deformations of such a continuum. In such a deformation, the co-ordinate of position of a material element before and after deformation do not coincide, i.e. if we wish to connect the initial and final positions of one and the same particle, we cannot avoid introducing an uncertainty associated with torsion such as [14] :

$$\Delta x^{\mu} = \int Q^{\mu}_{\alpha\beta}\ dx^{\alpha} \wedge dx^{\beta} \qquad (10.17)$$

over an area element defined by the direction vectors $dx^{\alpha} \wedge dx^{\beta}$. This uncertainty arising from the torsion is essentially due to the non-holonomic character (*) of the plastic

---

(*) Non-holonomic in general mean that when a quantum operator acts on classical or quantum system, the initial and final states of the system do not coincide. Here we mean that the plastic deformation of the space-time continuum, creates difference in the initial and final position of the test particles, so that there is a built-in uncertainty in position owing to non-closure of pairs.

deformation. All geometrically definable quantities like affine connection, torsion and curvature are functions of the initial co-ordinates. When related to the final co-ordinates after the deformation (x') they acquire a non-holonomic character, i.e. there is an intrinsic uncertainty associated with them that is related to the topological change affected by the deformation, while change in position is affected by torsion is a change in momentum affected by curvature?

For a sufficiently small area element dS, uncertainty in distance between initial and final position would be

$$\Delta l^{\mu} = Q^{\mu} dS \qquad (10.18)$$

and this would induce fluctuations in distance in the metric through $\Delta l = \sqrt{g_{\mu\nu} dx^{\mu} dx^{\nu}}$. So what is important is not the point themselves but the "fluctuations" in their position, i.e. the interval between them caused by a plastic deformation of space itself through torsion. With quantized values, these fluctuations would also manifest as metric fluctuations; torsion and curvature are both necessary for quantum effects in space-time: since curvature causes relative acceleration between neighbouring test particles through the relation [15]:

$$a^{\mu} = R^{\mu}{}_{\alpha\beta\gamma} \frac{dx^{\alpha}}{ds} \frac{dx^{\beta}}{ds} n^{\gamma} \qquad (10.19)$$

(here interpreted as relative acceleration between initial and final positions due to deformations) and acceleration measures rate of change of momentum, i.e. we have momentum uncertainty due to curvature:

$$ma^{\mu}dS = \Delta p^{\mu} = R^{\mu}{}_{\alpha\beta\gamma}dx^{\alpha}dx^{\beta}dx^{\gamma} = R\ dS\ n^{\mu} \qquad (10.20)$$

So as position fluctuations are given by torsion, momentum fluctuation are due to curvature and we can interpret quantum effects (and then uncertainty principle) as consequences of space-time deformation i.e.

$$\Delta p^{\mu}\cdot\Delta x = \hbar\ ;\ \text{where } \Delta x^{\mu} = Q\ dS\ n^{\mu} \text{ and } \Delta p^{\mu} = R\ dS\ n^{\mu} \qquad (10.21)$$

and Schrödinger wave function may be interpreted in the geometrical characteristics of underlying space-time.

In Sakharov's picture of space-time as an elastic continuum, the gravitational constant which occurs in the Einstein-Hilbert action is interpreted as the metrical elasticity of space-time.

In flat space-time the zero point energy of quantum fields is formally divergent. In a curved space-time, the zero point energy is modified due to the curvature and can be formally expanded in a series

$$L(0) \propto R\int k\ dk + (\ldots.)^2 + \ldots\ldots \qquad (10.22)$$

(k is the wave number)

The analogy of the R term to the Hilbert-Einstein action

$(c^3/16\ \pi G)R$ suggests the formal identification

$$16\ \pi G \ = \ (c^3/\hbar)\ \left(\int k dk\right)^{-1} \qquad (10.23)$$

where, in order to set a finite value for G, one must introduce arbitrarily a cut off corresponding to a maximum wave number k_{max} . In order to achieve numerical consistency with the Newtonian gravitational constant G_N , Sakharov assumes that k_{max} should be $k_{max} \sim (c^3/\hbar G)^{1/2}$ which is the inverse of the Planck length (k, the wave number, having the dimension of $(\text{length})^{-1}$. Thus the cut off has to be assumed in this picture. However considering space-time as in the above picture we have an elastic continuum undergoing a plastic deformation due to torsion when the gravitational effects of spin are also considered. So in the above expression for L(0), the curvature R and higher powers must be evaluated with the antisymmetric connection containing the torsion tensor.

The presence of the torsion would naturally introduce a cut-off given by Q, the quantization of the defect non closure naturally implying a $Q_{max} = (c^3/\hbar G)^{1/2}$ merely by having a spin $(1/2)\hbar$ as the source for the gravitational field.

Accordingly, the gravitational constant arises through

$$16\ \pi\ G_N \ = \ (c^3/\hbar)\ \left(\int_0^{Q_{max}} k dk\right)^{-1} \qquad (10.24)$$

As Q_{max} is inbuilt into the torsion structure of space-time once quantized spin is introduced, G_N is fixed. Thus introducing plastic deformation into Sakharov's pictures eliminates need for an arbitrary wave number cutoff to determine G_N . Thus the gravitational constant G_N arises from quantum effects of torsion on space-time.

References

[1] C.Sivaram and K.P.Sinha - Physics Reports 51,111 (1979)
[2] V.de Sabbata and C.Sivaram - Il Nuovo Cimento 105B, 603 (1990)
[3] V.de Sabbata and C.Sivaram - Astrophys.Space Sci.158, 347 (1989)
[4] V.de Sabbata and C.Sivaram - Astrophys.Space Sci.165, 51 (1990)
[5] C.Sivaram and V.de Sabbata - Ann.der Physik 48, 362 (1991)
[6] V.de Sabbata and C.Sivaram - Ann.der Physik 48, 419 (1991)
[7] Y.Fujiwara, S.Higuchi, A.Hosoya, T.Mishima and M.Siino - Class.Quantum Grav.9, 867 (1992)
[8] V.de Sabbata and C.Sivaram - "Torsion, Quantum Effects and the problem of cosmological constant" in vol. devoted to 75th jubilee of Peter Gabriel Bergmann - Vol.56, E.Majorana Int.Sci.Series ed.by Antonino Zichichi, Venzo de Sabbata and Norma Sanchez Plenum Press N.Y.1991, pag.19
[9] V.de Sabbata and C. Sivaram - Astrophys.Space Sci. 165, 51 (1990)
[10] V.de Sabbata and C.Sivaram - "Negative mass and Torsion" in vol.devoted to 80th jubilee of Ya.P.Terletski, Moscow ed. 1991, pag.79

[11] V.de Sabbata and C.Sivaram - Astrophys.Space Sci. 176, 141 (1991)
[12] V.de Sabbata, P.I.Pronin and C.Sivaram - Intern.J.Theor. Phys. 30, 167 (1991)
[13] Kawamura - Z.Phys. B29, 101 (1978) and B30, 1 (1978)
[14] V.de Sabbata, C.Sivaram, H.-H.v.Borzeszkowski and H.-J. Treder - Ann.der Physik 48, 197 (1991)
[15] C.Sivaram and V.de Sabbata - Astrophys.Space Sci. 176, 145 (1991).

CHAPTER XI

MISCELLANEA

1. Torsion and Wormholes

We shall first indicate how the antisymmetric field strength for torsion can give rise to instanton wormhole solutions. In the framework of an SL(2,C) gauge theory the Einstein-Cartan action can be expressed as [1,2,3]:

$$L = -\frac{1}{16\pi G} R + \frac{4}{8\pi G} S_{\alpha\beta\gamma} S^{\alpha\beta\gamma} \tag{11.1}$$

where the antisymmetric field strength $S_{\alpha\beta\gamma}$ is (related to the conserved current closing on the SL(2,C) algebra as

$$J^{\mu} \Longrightarrow J^{\mu} + \frac{1}{16\pi G} \varepsilon^{\mu\alpha\beta\gamma} S_{\alpha\beta\gamma}):$$

$$S_{\alpha\beta\gamma} = c_{\alpha} \times f_{\beta\gamma} \qquad \text{and} \qquad f_{\beta\gamma} \cdot g = F_{\beta\gamma}, \tag{11.2}$$

where $\bar{g} \equiv (g_1, g_2, g_3)$ are tangent vectors to the generators of SL(2,C) gauge group. $F_{\beta\gamma}$ in terms of the gauge fields A_{β} has the usual structure $F_{\beta\gamma} = \partial_{\beta} A_{\gamma} - \partial_{\gamma} A_{\beta} + [A_{\beta}, A_{\gamma}]$.

In the absence of matter, but with only torsion (analogous to neutrinos with spin but no mass), and with the usual Euclidean space-time metric ansatz: $ds^2 = d\tau^2 + a^2(\tau) d^2\Omega_3$, ($d^2\Omega_3$ being the metric of S^3), the field equations reduce to:

$$\left(\frac{da}{d\tau}\right)^2 = \left(1 - \frac{r^4_{min}}{a^4}\right), \tag{11.3}$$

where $r^4_{min} = 3G^2S^2/c^4$, S being he spin scalar. With S identified as the basic unit $\hbar$ of angular momentum, the minimum radius has the value of the Planck length and is not arbitrary as in the case of solutions involving other antisymmetric field strengths with random coupling (for instance we have found that in the case of black hole evaporation with torsion (see sections 4 and 5 of Chp.VIII and ref.[4]) effects, we are led at a final stage of a remnant mini black-hole with angular momentum of $\hbar$) so that we find the Coleman [5] result without any arbitrary constants. The time $\tau = 0$ is chosen, when radius attains minimum value. We have obtained a solution which describes tunneling from one E space time to another via a baby universe of radius r_{min} with a tunneling amplitude

$$\psi \sim k \exp(-S_{min}/\hbar) \qquad (11.4)$$

where

$$S_{min} = \hbar(3\pi^2/4)(r_{min}/l_{Pl}). \qquad (11.5)$$

With an effective cosmological constant $\Lambda(\alpha)$,

$$\Lambda(\alpha) = \Lambda_o - \alpha \ k \exp(-S_{min}/\hbar), \qquad (11.6)$$

we can arrive at Coleman's expression for the wave function of the universe

$$\psi = \int_{-\infty}^{+\infty} \frac{d\alpha}{(\pi)^{1/2}} < \exp(V_{min}\alpha)>^{\beta} Z(\alpha) , \qquad (11.7)$$

where V is the volume of the 3-space and the weight $Z(\alpha)$ is given by:

$$Z(\alpha) = \exp(-\alpha^2/2)\exp\left(k \exp\frac{3}{8G^2\Lambda(\alpha)}\right), \tag{11.8}$$

having an extremely sharp peak at $\Lambda(\alpha) = 0$.

These wormholes can be considered as describing the tunneling from one black hole state with spin to another (like the Einstein-Rosen bridge [6]), or transition from spinning black hole to string.

For example wormhole perturbed from its minimal equilibrium position at $a = r_{min}$ evolves

$$\ddot{a} = -(3G^2s^2)/2c^4a^5 \sim r^4_{min}/a^5 \tag{11.9}$$

or small deviation, imply maximal acceleration

$$\ddot{a} \sim c^{7/2}/(\hbar G)^{1/2} \tag{11.10}$$

2. Propagating torsion

i. When we consider propagating torsion it can be shown to be equivalent to a pseudoscalar field i.e.[7]

$$Q_\mu = \partial_\mu \phi \tag{11.11}$$

In this sense the measurable effects may resemble black holes with axionic charge. Gauge theory of torsion gives the

Einstein-Cartan action as:

$$L = \int d^4x\sqrt{-g}\,(-R/2\chi - (1/4\chi)S^{\mu\nu\beta}S_{\mu\nu\beta}) \qquad (11.12)$$

$$S_{\mu\nu\beta} = c_\mu \times f_{\nu\beta}\ ; \qquad f_{\nu\beta}\cdot\tilde{g} = F_{\nu\beta} \text{ etc.}$$

Through a duality transformation [8]:

$$L = \int d^4x\sqrt{-g}\ (-R/2\chi + (1/2)\partial^\mu\phi\ \partial_\mu\phi) \qquad (11.13)$$

Analogous to the axion case we have conservation of discrete charge through $\partial_\rho(\varepsilon^{\rho\mu\nu\sigma}S_{\mu\nu\sigma}) = J_A$.

ii. When considering forces governing the stability of black hole (i.e. not the field strength which is nonvanishing, but the vector potential), we have Aharanov-Bohm (A B) analogy [9].

The 3-form field strength can be expressed as curl of 2-form potential $H = dB$ and $Q = (1/2\pi)\int_\Omega H = (1/2\pi)\int_\Sigma B$ where Σ is the boundary of Ω. The equivalent (A B) phase is $\exp(i\int_\Sigma B = \exp(2\pi iQ)$. The world sheet of a torsion induced string defect lies on a closed surface Σ where the string carries a flux Φ. This is the (A B) phase acquired by the defect. This is similar to (A B) interaction of a magnetic solenoid with an electrically charged particle, i.e. we have *torsion* $\Longrightarrow$ *magnetic field* analogy ($Q\cdot S \Longrightarrow \mu\cdot B$) (see Chp.II).

We have also similarity to London flux equation for

penetration giving effective photon mass; a analogous to Z(N) charge "quantum hair" of black holes.

iii. There are also possible electromagnetic effects of spin-torsion coupling in rotating black holes. In fact the spin-torsion coupling arising from axial vector part of torsion tensor i.e.

$$A_i = (1/3!)\sqrt{-g}\ \varepsilon_{ijkl}Q^{jkl} \qquad (11.14)$$

interacting with virtual pairs gives these effects: the interaction has the form [7]:

$$L = (\alpha/16\pi)\int d^4x\sqrt{-g}\ \varepsilon^{ijkl}\ A_i F_{jk}\partial_l\phi \qquad (11.15)$$

giving the torsion modified Maxwell equations

$$(\partial/\partial x^{\nu})\left(\sqrt{-g}\ F^{\mu\nu}\right) = -(1/2)\alpha\varepsilon^{\mu\nu\alpha\beta}F_{\varphi\alpha}\partial_{\beta}\phi \qquad (11.16)$$

where $\Box\phi = (G\alpha/c^5)\varepsilon^{ijkl}F_{ij}F_{kl}$

Then we have corresponding dispersion relation [10]:

$$\omega^2/c^2 - k^2 = \pm\ 4\ \sqrt{(2/3c)}\ (ac^6\alpha^{3/2}/G^2M^2) \qquad (11.17)$$

(where a is the Kerr parameter and a R_S/r^3 is Lense-Thirring precession).

In case of maximal rotation for $M \approx 1M_{\odot}$, we get $\Delta\nu/\nu$ for

γ-rays $\approx 10^{-8}$. This may be detectable by Mössbauer technique. This is a higher order effect involving G^2 so suppressed by factor of G.

Moreover torsion gravimagnetic effect is comparable to Lense-Thirring effect for particle orbits around black holes ($Q = aR_S/r^3$, Q being the trace of torsion tensor); such effect must be considered for accretion discs around black holes.

3. <u>Torsion and strings</u>

In string theories, the coupling is characterized by the so called string tension T, which is the energy or mass per unit length. At the Planck scale ($M_{Pl} = (\hbar c/G_N)^{1/2}$), T is given by [11]:

$$T_{Pl} = c\ /G_N \simeq 1.6 \cdot 10\ \ g\ cm \qquad (11.18)$$

(G_N is the Newtonian gravitational constant).

Thus T has units of G^{-1} ($c = 1$), i.e. $T \propto 1/G$. At any other energy scale, M, T is given as

$$T = T_{Pl}(M/M_{Pl})^2 \qquad (11.19)$$

T scales as the mass squared, i.e. the energy at which the symmetry breaking occurs. So the maximum value of the string tension would be at the Planck mass, i.e. $T_{Pl} = c^2/G = T_{max}(M \leq M_{Pl})$. The string length at this scale is the Planck length $L_{Pl}=(\hbar G/c^3)^{1/2}$ so that the string mass is $T_{Pl}L_{Pl}$ which is just M_{Pl}!

In the usual picture it is not very clear how the string tension arises (for instance at the Planck scale) from the interaction between point particles. Here we explore its possible connection with the spin-torsion interaction between spinning constituents and to account for its scaling with energy. In Chapter I we had obtained the electron charge and radius from a consideration of the interaction between strong gravity and spin-torsion which may take place between spinning constituents.

Now as is known (see Chp.II eq.(2.10)) the torsion tensor $Q_{ij}{}^{k}$ is related to the spin density tensor $J_{ij}{}^{k}$ (which acts as the source term for torsion, analogous to matter density being source term for curvature) through the equation:

$$Q_{ij}{}^{k} = (8\pi G/c^2)(J_{ij}{}^{k} - (1/2)\delta^k_i J_{bj}{}^{b} - (1/2)\delta^k_j J_{ib}{}^{b}) \quad (11.20)$$

Thus for N aligned spins per unit volume, the torsion is given as [12]:

$$\bar{Q} = (4\pi G/c^2)\bar{J}, \quad (11.21)$$

where $J = N\hbar/2$

Consider now a string of Planck length $L_{Pl}= (\hbar G/c^3)^{1/2}$ with the interacting spins having a net spin of $\hbar/2$. Then the spin density is $J = (\hbar/2)/[(\hbar G/c^3)^{3/2}(4\pi/3)]$. The spin-torsion energy is then $Q\cdot\hbar/2$ and is thus given by:

$$E_{ST} = (4\pi G/c^2)(\hbar/2)(\hbar/2)/[(4\pi/3)(\hbar G/c^3)^{3/2}] \qquad (11.22)$$

Assuming the energy of the string is solely given by this spin-torsion field, then the energy or mass per unit length (i.e. the tension) is given by:

$$T = E_{ST}/L_{Pl}c^2 = (3/4)\left[\hbar^2/(\hbar G/c^3)^2\right](4\pi G/c^4) \simeq c^2/G \qquad (11.23)$$

which is just the tension in the string which is assumed in the superstring models with unification at the Planck scale. This shows the underlying importance of spin or torsion in determining the tension.

If the string length were now $L_{Pl}\cdot(M_{Pl}/M)$, i.e. the energy scale at which the strings are produced is M, $(M < M_{Pl})$, then the tension would evidently be reduced by the ratio of $(M/M_{Pl})^2$, i.e. it would now be:

$$T = (c^2/G)(M/M_{Pl})^2 = T_{Pl}(M/M_{Pl})^2 \qquad (11.24)$$

in agreement with eq.(11.19). This is also consistent with what we have seen in Chapter VI (see also [13,14]) wherein it was shown that the spin-torsion coupling constant G depends on energy in the early universe as E^{-2}, i.e. as M^{-2}, i.e. it can pass through the values of the coupling constants of the fundamental interactions during the evolution of the universe, enabling the understanding of the ratios of the different coupling strengths. Thus we had the relations:

$$G_N = G_F(M_w/M_{Pl})^2 \qquad (11.25)$$

(where G_F and M_w are the Fermi weak interaction constant and intermediate weak boson mass) and also:

$$(G_f/G_N) = (M_{Pl}/M_f)^2 \approx 10^{38} \qquad (11.26)$$

and

$$(G_f/G_F) = (M_w/M_f) \qquad (11.27)$$

where G_f is the strong gravity constant characterizing the strength of strong interactions, and $M_f \approx 1$ Gev is the f-meson mass. Eqs. (11.25) and (11.26) enable us to understand the coupling strengths of strong, weak and gravitational interactions. Again at the Grand unification scale where the interactions are mediated by exchanges of super-heavy gauge bosons of mass M_x, the corresponding coupling is:

$$G_{GUT} \simeq G_F(M_w/M_x)^2 \simeq G_N(M_{Pl}/M_x)^2 \qquad (11.28)$$

i.e. at Planck scale $G = G_N$.

The above relations would enable us to write the string tension as:

$$T = c^2/G_{eff} \qquad (11.29)$$

where G_{eff} is the value of G at the scale M,

$$G_{eff} = G_N(M_{Pl}/M)^2 \qquad (11.30)$$

so that (11.29) and (11.30) imply eq.(11.24).

For the hadronic string describing the Regge trajectories of hadrons, the tension is given by the appropriate strong gravity constant G_f, i.e.:

$$T_{strong} = c^2/G_f \simeq 10^{-11} g\ cm^{-1} \simeq 10^{-38} c^2/G_N \qquad (11.31)$$

The slope of the Regge trajectories is given by [15]

$$\alpha_R = (G_f/\hbar c) \simeq 1\ (Gev)^{-2} \qquad (11.32)$$

This is consistent with the observed mass spectroscopy of a large number of hadronic resonances states lying on these trajectories; i.e. we have a spectrum of the type $J(M) = \alpha_R M^2$, which can also be interpreted as the spectrum of the rotational states of a string with the tension related to α_R as $\alpha_R = (c/\hbar)(2\pi T)^{-1}$. The Hagedorn limiting temperature (for strongly interacting particles) is again related to α_R as $T \propto (1/\alpha_R)^{1/2} \simeq (1/K_B)(\hbar c^5/G_f)^{1/2} \simeq 10^{13}\ {}^oK$. For the superstring at the Planck scale, this is just $(1/K_B)(\hbar c^5/G_N)^{1/2} \simeq 10^{32} K$, the Planck temperature. We can now estimate the tension for strings generated at various energy scales (from eqs. 11.24, 11.29, 11.30). Some examples:

· for Grand unification strings or GUT strings,

$$M_x \approx 10^{15}\ \text{Gev},\quad T_{GUT} \simeq 2\cdot 10^{21} g\ cm^{-1} \simeq 2\cdot 10^{-7} c^2/G \qquad (11.33)$$

· cosmic strings with this tension are invoked for galaxy formation

· for supersymmetric grand unified strings, or Susy GUT strings:

$$T_{SusyGUT}(M_x \simeq 10^{16}\text{Gev}) \simeq 10^{24} g\ cm^{-1} \simeq 10^{-4} c^2/G \qquad (11.34)$$

· for SO(10) GUTs strings (symmetry breaking at 10 to 10Gev):

$$T_{SO(10)} \simeq 10^{-3} \text{to } 10^{-5}\ c^2/G \qquad (11.35)$$

· for electroweak strings (characterized by weak Fermi constant):

$$T_{e-w}(M_x \simeq 10^2\text{Gev}) \simeq 2\cdot 10^{-33}\ c^2/G \qquad (11.36)$$

The rotational states of all these strings would follow a spectrum of type $J(M) \simeq (G_{eff}/\hbar c)M^2$.

Interestingly, also astronomical objects show an angular momentum mass relation of type (see Chp.II eq.(2.2)):

$$J = pM^2 \qquad (11.37)$$

This again can be explained by torsion [16], [17] and the tension here has the form $T_{astron} = \alpha_{em} c^2/G$ is the electromagnetic fine structure constant. In short we have the possibility of a unified understanding of string tensions at all energy scales! (incorporating all interactions) through an effective spin-torsion gravitational interaction.

4. On limiting field strengths in gravitation

4a. Introduction

In Chp.VIII we have noted that the maximal acceleration exists and is given by $a_{max} = c^{7/2}/G^{1/2}\hbar^{1/2}$. This originates as a quantum effect due to torsion [18], and would give rise to a maximal force, as we shall explain, which can be operationally defined as $F_{max} = c^4/G$.

Briefly we remember (see Chp.II eq.(2.12))that the spin-torsion interaction energy between spin S and torsion Q has been shown to be [19]

$$E = - \bar{S}\cdot\bar{Q} \qquad (11.38)$$

and the formal analogy of this expression with the interaction energy of a magnetic dipole $\bar{\mu}$ in constant magnetic field $\bar{H}$ i.e. $\bar{\mu}\cdot\bar{H}$, shows that spin in spaces with torsion behaves like a dipole in a magnetic field, i.e. will orient predominantly along the torsion direction. Then it is easy to show that we are led to a magnetic field

$$B = (8\pi/3c)(2\alpha G)^{1/2}\sigma \qquad (11.39)$$

where α is the fine structure constant and σ the spin density. In that way [20] we obtain a maximum primeval field at the Plank epoch of 10^{58}G.

Subsequently flux conservation would have made the field behave as $Bt = BT^{-2} = $ const. with expansion time t and temperature T. Such a magnetic field would have accelerated charged particles in the early universe. For a particle with charge e, the relativistic Larmor frequency is given by $\omega_L = (ecB/\hbar)^{1/2}$ implying a magnetic energy $\hbar\omega \sim (e\hbar cB)^{1/2}$. It is known that this would imply a critical magnetic field when $(e\hbar cB)^{1/2}$ equals the rest mass energy mc^2 of the particles - i.e. when the gyroradius r/G becomes smaller than the Compton length. Thus quantum considerations impose a critical magnetic field strength of $B_c = m^2c^3/e\hbar$. At the Planck epoch when $m = m_{Pl} = (\hbar c/G)^{1/2}$, this implies a B_{max} of $c^4/eG \approx 10^{58}$G; now, as $\omega_L = (ecB/\hbar)^{1/2}$ implies a circular (or helical) acceleration in the magnetic field of the charged particle with $a \cong \omega_L^2 r_G$ we will have $a \sim (ecB/\hbar)r_G$. If we substitute $B_{max} \approx c^4/Ge$ and the corresponding $r_G = (\hbar G/c^3)^{1/2}$ at the Planck epoch we have the expression for maximal acceleration as given before i.e. $a_{max} = c^{7/2}/(\hbar G)^{1/2} = m_{Pl}c^3/\hbar = 5\cdot 10^{53}\text{cm s}^{-2}$. So this expression for maximal acceleration is originated as a quantum effect due to torsion and one can also note that in this expression of maximal acceleration for a charged particle the electric charge is not involved : instead there is the Planck constant which is connected with spin.

The minimal length L_{min}, (see Chp.VIII) also dictated by torsion [21], implies a maximal mass or energy.

Also here one can remember that this minimal length can be arrived at by considering the non-closure property of a contour in the space-time with torsion, that is:

$$l^{\alpha} = \oint Q^{\alpha}_{\mu\nu}\, dA^{\mu\nu} \neq 0 \qquad [11.40]$$

(where $dA^{\mu\nu} = dx^{\mu} \wedge dx^{\nu}$ is the area element enclosed by the loop) over a closed infinitesimal contour is different from zero; this non closure property can be treated as defects in space-time in analogy to the geometrical description of dislocations (defects) in crystals, and this can constitute a way to go toward the quantization of gravity. In fact we have found (see Chp.VIII) that as torsion is related to the intrinsic spin, if we connect torsion to the fundamental unit of intrinsic spin $\hbar$, we find that the defect in space-time topology should occur in multiplies of the Planck length, so that we can write $\oint Q\, dA = n(\hbar G/c^3)^{1/2}$ and for the same reason considering the fourth component we have also an analogous situation as regards the time, namely $t = (1/c)\oint Q\, dA = n(\hbar G/c^5)^{1/2}$ which gives a minimum unit of time $\neq 0$ (for $n = 1$). So the minimal length and the minimal time are dictated by torsion; this implies a maximal mass or energy. For a quantum state we have $M_{max} = (\hbar/L_{min}c) = (\hbar c/G)^{1/2}$. This would give rise to a maximal force as:

$$F_{max} = M_{max}\, a_{max} = c^4/G \tag{11.41}$$

This expression can be also be arrived at by noting that given a mass M, general relativistic considerations imply that it cannot be localizedto a distance R_{min} less than G M/c^2 (for the Schwarzschild case it is $2MG/c^2$ and for maximal rotation [of the type $J = GM^2/c$ which we have earlier shown (see Chp.VIII and ref.[4] is to be expected for object forming black holes) it is GM/c^2).

The gravitational self force $F = GM^2/R^2$ is thus maximized for R_{min} giving $F_{max} = GM^2/R^2_{min} = c^4/G$, that is (11.41). Note that F_{max} is independent of M showing that it is a universal quantity. Then the maximal mass operationally definible is F_{max}/a_{min}.

We shall now indicate how a_{min} can be defined: as clearly shown in ref.[5], we have the formula for the acceleration due to torsion in an exact solution of Einstein-Cartan theory as

$$\ddot{R} = G^2S^2/c^4R^5 \tag{11.42}$$

We had applied this formula in the case of black hole evaporation. Here we shall indicate how the above relation can be used to obtain both a_{max} and a_{min} . For a_{max} using $S \approx \hbar$ and $R_{min} = (\hbar G/c^3)^{1/2}$ (as explained earlier), we get

$$\ddot{R}\,)_{max} = a_{max} = G^2\hbar^2/c^4(\hbar G/c^3)^{5/2} = c^{7/2}/(\hbar G)^{1/2} \tag{11.43}$$

which agrees with what was stated above!

For a_{min} we have to invoke cosmological parameters in the above equation. For S we take the total spin of universe which, as was shown earlier (see eq.(5.42) and ref.[22]), was given by a cosmological solution involving torsion as:

$$S_U = 10^{120}\hbar \tag{11.44}$$

which for R, the maximal radius i.e. $R = R_H$, the Hubble radius, we have

$$\begin{aligned}\ddot{R} = a_{min} &= G^2S_U^2/c^4R_H^5 = \left(c^{7/2}/(\hbar G)^{1/2}\right)\left(S_U/\hbar\right)^2\left(L_{Pl}/R_H\right)^5 \\ &= a_{max}\left(S_U/\hbar\right)^2\left(L_{Pl}/R_H\right)^5 \simeq 10^{-8}\text{cm sec}^{-2}\end{aligned} \tag{11.45}$$

which agrees with what is assumed in MOND (see section 4c).

So we see that both a_{max} and a_{min} follow as consequence of an exact solution of Einstein-Cartan theory based on torsion.

We will now proceed to estimate the maximal and minimal temperature associated with a_{max} and a_{min} respectively. First of all, as we have shown [23], we notice that the minimal operationally definable temperature is (see eq.(8.31)):

$$T_{min} = \hbar c/k_B R_H \tag{11.46}$$

where R_H , the maximal background scale, is the Hubble radius.

Now we shall proceed to show that this minimal temperature follow from the minimal acceleration a_{min} as given above by the effect of torsion. In fact another familiar result that an observer in a state of uniform acceleration, a, finds himself in a thermal bath of temperature [4]

$$T = \hbar a/k_B c \qquad (11.47)$$

Thus a_{min} as given above implies a minimal temperature of

$$T_{min} = \hbar a_{min}/k_B c \qquad (11.48)$$

Substituting for a_{min} as found above through torsion, we find finally (after substituting also for S_U etc.)

$$T_{min} = \hbar c/k_B R_H \qquad (11.49)$$

which is the same as equation (11.46). From eq.(11.48) we can express a_{min} as

$$a_{min} = k_B T_{min} c/\hbar \qquad (11.50)$$

Substituting equation (11.46) for T_{min} and using $R_H = c/H_0$ then gives

$$a_{min} = c\ H_0 \qquad (11.51)$$

Then the maximal mass operationally definible is F_{max}/a_{min} i.e.

$$M_{max} = c^4/G\,c\,H_0 = c^3/G\,H_0 \qquad (11.52)$$

For $H_0 \approx 50$ km/(s Mpc), this gives

$$M_{max} = 3\cdot 10^{56}\text{g} \qquad (11.53)$$

Usually H_0 is associated with the critical density ρ_c as

$$\rho_c = 3\,H_0^2\,/8\pi\,G \qquad (11.54)$$

(for H_0 = 50 km/(s Mpc), $\rho_c \approx 5\cdot 10^{-30}$g cm^{-3})

Here on the contrary, we relate maximal mass to H_0 through eq.(11.52). However eq.(11.54) does not tell us the actual density of the universe. For this we need to know the deceleration parameter q given by

$$q = -\ddot{R}/(RH_0^2) \qquad (11.55)$$

Using for $\ddot{R}$ the a_{min} that is $\ddot{R} = c\,H_0$ and $R = c/H_0$, we get $q = 1$, and since $q = \Omega/2$ (where $\Omega = \rho/\rho_c$) this gives $\Omega = 2$ (for the maximal mass) i.e. ρ, the actual density, is $2\,\rho_c \sim 10^{-29}$g cm^{-3}. This is the density corresponding to the maximal mass. So just from a_{min} and F_{max} we are able to give definitive values for M_{max} and ρ for the universe. As $M_{Pl} = 2\cdot 10^{-5}$g we have $M_{max}/M_{Pl} = 1.5\cdot 10^{61}$ and $t_{max}/t_{Pl} = 1.5\cdot 10^{61}$, and using for t_{Pl} the value $t_{Pl} = 5\cdot 10^{-44}$s ($L_{Pl}/c = 1.6\cdot 10^{-33}/3\cdot 10^{10}$) we have for the maximal age of universe

$$t_{max} \approx 1.5\cdot10^{61}\cdot5\cdot10^{-44}s =$$
$$= 7.5\cdot10^{17}s \sim 2.3\cdot10^{10}yr \cong 23 \text{ billion yr} \qquad (11.56)$$

Thus in this way of defining mass directly through H_0 rather than in the usual method of obtaining density from H_0 and then multiplying by volume, we have no need for postulating a huge amount of dark matter.

4b Maximal gravitational field strength

The above F_{max} enables us to write for the maximal gravitational field strength E for a given mass M as:

$$E_{max} = F_{max}/M = c^4/G\,M \qquad (11.57)$$

As $M_{min} = M_{Pl} \simeq (\hbar c/G)^{1/2}$, we have for the absolute maximal gravitational field strength the quantity

$$E_{max} = c^4/[G(\hbar c/G)^{1/2}] = \left(c^7/G\,\hbar\right)^{1/2} \qquad (11.58)$$

which is the same as the maximal acceleration a_{max}. Thus this establishes the equivalence between maximal acceleration and maximal gravitational field strength. The existence of such a maximal field strength for gravity is analogous to the maximal field in Born-Infeld electrodynamics. This model gives a finite size for any gravitating collapsing object, just as Born-Infeld electrodynamics gives finite value for the electron self energy(#). Thus in this case, just as in the Born-Infeld case, the electron finite radius is defined in terms of E_{max} as $r^2 = e/E_{max}$, and in the case of a gravitating body, the existence of a <u>universal</u> $E_{max} = \left(c^7/G\hbar\right)^{1/2}$, would imply that for a gravitating body of mass M, the minimum size would be given by:

$$R^2_{min} \simeq G\ M/E_{max} \tag{11.59}$$

or

$$R_{min} \simeq (G\ M/E_{max})^{1/2} \tag{11.60}$$

i.e.

$$R_{min} - \left(G\ M/(c^7/G\hbar)^{1/2}\right)^{1/2} = \left(G^3\hbar/c^7\right)^{1/4} M^{1/2} \tag{11.61}$$

For instance in the case we are considering for M the Planck mass, we find

$$R_{min} = (\hbar\ G/c^3)^{1/2} \tag{11.62}$$

that is the Planck length.

(#) *Note that in our case such results, both for electromagnetic and gravitational field, are obtained when torsion is considered [24]: in fact we are arrived at a minimum radius, for instance, of the electron, (see Chp.VII eq.(7.13)) starting explicity from a modified Poisson's equation when a term containing the spin is incorporated through torsion.*

So the existence of a Maximal gravitational field strength implies that any gravitating body with a given mass M, cannot have a radius smaller than R_{min} given by eq.(11.61). The presence of $\hbar$ shows that this is a quantum effect; for the classical limit $\hbar \Longrightarrow 0$, $R_{min} \Longrightarrow 0$, giving singular value as in usual case.

For the universe with $M = 3\cdot 10^{56}$g (see eq.(11.53)), this would give $R_{min} = 10^{-2}$cm. Note that Trautman [25] finds 1 cm. However in the case of Trautman, R_{min} occurs only if all spins are aligned. Here it is a consequence of the existence of a maximal gravitational field strength and is independent of any specific physical situation. It is a universal lower limit for R (if the gravitational field strength is ∞ as in the classical case $\hbar \Longrightarrow 0$ then $R \Longrightarrow 0$, i.e. we have the usual point singularity).

For solar mass, $R_{min} \approx 10^{-14}$cm, i.e. a black hole cannot collapse even inside the Schwarzschild sphere to R = 0 (as in the conventional picture) but must halt at $R_{min} \sim 10^{-14}$cm (because this is governed by universal maximal field strength).

For a Planck mass black hole $R_{min} = 10^{-33}$cm. For the mass of universe, being $R_{min} = 10^{-2}$cm and mass $\approx 3\cdot 10^{56}$g we have $\rho_{max} = 10^{63}$g cm^{-3}. The same result we would have obtained considering that ρ scales as $M^{-1/2}$ i.e $\rho_{max} = \rho_{final} = \rho_{Pl}(M_{Pl}/M)^{1/2}$, so that for universe we have

$$\rho_{max} = (c^5/G^2\hbar)(M_{Pl}/M)^{1/2} = 10^{63}g\ cm^{-3} \qquad (11.63)$$

(Trautman found $\rho_{max} = 10^{55}g\ cm^{-3}$).

This is the absolute maximal possible density for the collapsed universe irrespective of any physical situations like spins alignment, magnetic field etc.. For solar mass black hole:

$$\rho_{max} = 10^{93}(M_{Pl}/M_{\odot})^{1/2} \simeq 10^{74}g\ cm^{-3} \qquad (11.64)$$

Thus even inside its Schwarzschild sphere, a solar mass black hole cannot collapse to a density higher than that given by (11.56) (in the conventional picture a comoving observer also collapses to infinite density inside the Schwarzschild sphere).

So all these are interesting consequences of the existence of a maximal gravitational field strength.

4c A basis for the fundamental acceleration in MOND

As regards the minimal acceleration, $a_{min} = cH_0$ (see eq.(11.51)) we note that it is the same value as given in the so called MOND (Modification of Newtonian Dynamics) [26],[27] which has had considerable phenomenological success in explaining a wide variety of observations involving galaxies and clusters such as well known relationships involving a luminosity-rotation velocity or luminosity-velocity dispersion etc..

In fact this theory requires empirically $a_0 \approx c H_0$, where H_0 is the Hubble constant. In the above approaches, despite the phenomenological success, the introduction of a_0 is ad hoc and this is a common complaint about the theory [26,27]. Is there any deeper fundamental basis for the existence of $a_0 \approx c H_0$?

As the authors of ref.[27] point out, there could be a profound connection with cosmology which would render the theory less ad hoc. Also despite its success at the phenomenological level, it could be made more respectable by connecting it to other familiar physical laws.

Thus the existence of a fundamental minimal acceleration or temperature can be considered a consequence of the fact that background cosmological space is curved. In a flat background space $R_H \Longrightarrow \infty$, $T = 0$, and $a_0 = 0$.

A curved background space is thermodynamically associated with a fundamental temperature (following Hawking-Unruh) and corresponding to this is a fundamental acceleration. It comes out nicely as $a_0 = c H_0$, which is what is required by MOND. So the existence of such a minimal acceleration (purely a consequence of background curvature), means that the velocities in the outermost parts of halos (where acceleration approach a_0) cannot fall with distance in a Keplerian way, but must tend to a constant value with distance.

Thus in this picture, there is no need for vast amounts of dark matter to explain rotation curves of galaxies. In fact in the way of defining mass directly through eq.(11.52) rather than the usual method of obtaining it through density from H_0 , we have no need to postulate dark matter in order to close the universe.

Finally we note that the existence of a minimal acceleration would imply, again from the equivalence principle, a minimal field strength. This would have interesting consequences not only for large systems like galaxies but also for planetary and stellar systems. Thus given an object of mass $M_\odot$, we can estimate the maximum radius it can have and be still gravitationally bound, i.e.:

$$R_{max} = (G M_\odot / a_{min})^{1/2} \qquad (11.65)$$

This gives for $M = M_\odot$

$$R_{max} \approx 3 \cdot 10 \quad cm \qquad (11.66)$$

This is about 10 times the size of the solar system. For red giants, $R \approx 10^{15}$cm. This corresponds to a minimum density of

$$\rho_{min} \approx \left(a_{min}^{3/2} / G^{3/2} M_\odot^{1/2} \right) \qquad (11.67)$$

This is much greater than the interstellar density of ~ 10^{-23}g cm^{-3} So we can say this is the minimal density for a system of mass $M_\odot$ to be gravitationally bound and begin collapsing. This is independent of other physical considerations, from thermodynamics, hydrodynamics etc. For the universe as a whole, we have a $R_{max} \approx 1.5\cdot 10^{28}$cm. Perhaps this would be the maximum radius to which the universe with M_{max} would expand.

So the existence of maximal and minimal gravitational field strength has several implications.

References

[1] C.Sivaram and K.Sinha - Physics Reports 51, 111 (1979)

[2] M.Carmeli - Group Theory and G.R.,McGraw Hill N.Y.(1984)

[3] P.Bandyopadhyay - preprint, Calcutta (1990)

[4] V.de Sabbata and C.Sivaram - "The central role of spin in black hole evaporation" in 'Black hole physics' Kluwer Academic Publ. NATO ASI Series, Vol.364, 225 (1992)

[5] S.Coleman - Nucl.Phys. B307, 867 (1988)

[6] A.Einstein and N.Rosen - Phys.Rev.48, 73 (1935)

[7] V.de Sabbata and M.Gasperini - Phys.Rev.23D, 2116 (1981)

[8] V.de Sabbata and C.Sivaram - Il Nuovo Cimento 105B, 1181 (1990)

[9] V.de Sabbata, Pronin and C.Sivaram - Int.Journ.Theor. Phys.30, 1671 (1991)

[10] V.de Sabbata and M.Gasperini - Nuovo cimento 28, 181 (1980) and Nukleonika 25, 1373 (1980)

[11] V.de Sabbata and C.Sivaram - Ann.der Physik 48, 419 (1991)

[12] V.de Sabbata and M.Gasperini - Nuovo Cimento 27, 289 (1980)

[13] V.de Sabbata and C.Sivaram - Astrophys.Space Sci.158, 347 (1989)

[14] V.de Sabbata and C.Sivaram - Astrophys.Space.Sci.165, 51 (1990)

[15] V.de Sabbata and C.Sivaram - Ann.der Physik 48, 362 (1991)
[16] P.Brosche - in "Spin, Torsion, Rotation and Supergravity", Peter G.Bergmann and Venzo de Sabbata eds.,Plenum Press New York 1980 p.375
[17] V.de Sabbata - in "The Origin and Evolution of Galaxies" ed.by Venzo de Sabbata, World Scientific Singapore 1982 p.181
[18] V.de Sabbata and C.Sivaram - Astrophys.Space Sci.176, 145 (1991)
[19] V.de Sabbata and M.Gasperini - Lett.Nuovo Cimento 27, 138 (1980)
[20] V.de Sabbata and C.Sivaram - Nuovo cimento 102B, 107 (1988)
[21] V.de Sabbata, C.Sivaram, H.-H.v.Borzseszkowski and H.-J.Treder - Ann.der Physik 48, 197 (1991)
[22] V.de Sabbata - "On the relation between fundamental constants" in 'Gravitational Measurements, Fundamental Metrology and Constants' Kluwer Academic Publ. NATO ASI Series, Vol 230, 115 (1989); see also V.de Sabbata and M.Gasperini - Lett.Nuovo Cimento 25, 489 (1979)
[23] V.de Sabbata and C.Sivaram - Found.Phys.Lett.5, 183 (1992)
[24] V.de Sabbata and C.Sivaram - Ann.der Physik 47, 511 (1990)
[25] A.Trautman - Nature 242, 7 (1973)
[26] D.Lindley - Nature 359, 583 (1992)
[27] M.Milgrom and R.H.Sanders - Nature 362, 25 (1993)

CHAPTER XII

OTHER PHYSICAL CONSEQUENCES OF TORSION

1 Neutron Interferometry

1a. Introduction

A large number of theoretical models and experiments claim the existence of new intermediate range forces. Results of the experiments [1 - 10] and phenomenological considerations suggest that there might be a deviation from the Newtonian inverse square law with a potential of the form

$$V(r) = - G_{\infty}(M/r)\{1 + \Delta V\} \qquad (12.1)$$

Here $\Delta V(r)$ is caused by existence of a new type of interaction which was called the "fifth force".

Evidently there are several questions. The most important of them are:

a) if the 'fifth force' exists, to what attribute of matter could it couple?

b) what is (are) the mediating particle (particles) of the 'fifth interaction'?

c) what types of new effects both in classical and quantum regions can be predicted in the presence of the new interaction?

At present time there are a set of different opinions on the possible existence and nature of 'fifth force', and there are numerous classes of unified field models in which the new types of interactions can be predicted. In this short paper we consider the quantum treatment of the fifth force due to the exchange of tordions in the frame of gauge gravitational theory, and regard the two possible ways for experimental verification of this model of the 'fifth interaction'. In the next section we will give a phenomenological description of the neutron interferometry scheme. In the third section we propose models for fifth force mediators, i.e.mediating particles. Finally in the last section we consider the neutron interferometry effect due to interaction of neutrons with 'tordion field'.

1b. Fifth force and neutron interferometry

Fifth force is usually assumed to have a potential of the form (modification of the Newton law)

$$V(r) = -G_\infty(M/r)\{1 + \alpha \exp(-r/\lambda)\} \qquad (12.2)$$

The potential energy due to gravity and an extra Yukawa interaction between two macroscopic bodies is given by:

$$V_E(r) = -\int dr_1 \int dr_2 [G_\infty \rho(r_1)\rho(r_2)/r_{12}][1 + \alpha \exp(-r_{12}/\lambda)] \qquad (12.3)$$

where ρ_1 and ρ_2 are mass densities, α is the strength of coupling of the fifth force and λ is the range.

The energy contribution of the fifth force to the potential energy of a single neutron in the Earth's gravitational field is given by integration over the distribution of mass density acting on the neutron as:

$$V_F(r) = (G_\infty E_n \alpha/c^2)\int d^3r' \left(\rho(r')/|\bar{r} - \bar{r}'|\right) \exp\{- |\bar{r} - \bar{r}'|/\lambda\} \tag{12.4}$$

here E_n is the neutron energy. For non relativistic neutrons $E_n/c^2 = m_n$ is the neutron rest mass. Integrating over the Earth volume and approximately assuming a uniform density $\rho = 3M_\oplus/4\pi R_\oplus^3$ where $M_\oplus$ and $R_\oplus$ are respectively the mass and radius of the Earth.

Assuming $\lambda/R_\oplus \ll 1$ (as $\lambda \approx 200\text{m} \div 1\text{km} \ll R_\oplus \simeq 10^4\text{km}$), the above formula gives the additional potential energy contribution due to the fifth force acting on a neutron as (a neutron at the Earth's surface $r \simeq R_\oplus$)

$$V_F(R_\oplus) = (E_n/c^2)\alpha\ (G_\infty M_\oplus/R_\oplus)(\lambda/R_\oplus)^2 \tag{12.5}$$

So for neutrons of different energies E_{n_1}, E_{n_2}, the energy difference is:

$$\Delta V_F(R_\oplus) = (E_{n_1} - E_{n_2}/c^2)\alpha\ (GM_\oplus/R_\oplus)(\lambda/R_\oplus)^2 \tag{12.6}$$

For 20 Gev n's and 10 Gev n's we find, with $\alpha \simeq 10^{-3}$, $\lambda \simeq 200\text{m}$, $R_\oplus \simeq 6.4\cdot 10^8\text{cm}$, $\alpha(\lambda/R_\oplus)^2 \simeq 9\cdot 10^{-13} \simeq 10^{-12}$:

$$\Delta V(R_\oplus) \approx 10^{-14}\text{ev} \tag{12.7}$$

For neutron at rest or non relativistic neutrons

$$V_F(R_\oplus) \approx 10^{-3}\alpha(\lambda/R_\oplus)^2 \approx 10^{-15}ev \qquad (12.8)$$

Now let us consider the phase shift on a neutron in a neutron interferometer. Gravitational phase shift was found in COW experiment [10] and it is

$$|\Delta\Phi_G| = 4\pi\ (m/\hbar)(g/2v)A \qquad (12.9)$$

where m is gravitational mass, g the acceleration due to gravity, v is velocity of neutrons moving in field and A is area enclosed by interferometer beam paths. The explicit expression for $|\Delta\Phi_G|$ will be

$$|\Delta\Phi_G| = |\Delta\Phi_0|\{1 + (1/2)(gH/c^2)\} \qquad (12.10)$$

where

$$|\Delta\Phi_0| = 4\pi(m_0/\hbar)(g/2v)A \qquad (12.11)$$

(with m_0 the rest mass of neutron)
and gH/c^2 is the gravitational potential energy and H is height difference of paths. In general

$$m = m_0\{1 + 2gH/c^2\}(1 + 2gH/c^2 - v^2/c^2)^{-1/2} \qquad (12.12)$$

and we can consider the relativistic effects.

In the fifth force presence we have additional

contribution to potential energy of neutron, i.e. to gH/c^2 i.e.

$$gH/c^2 + V_F(R_\oplus) \quad \text{or} \quad \Delta V_F(R_\oplus) \tag{12.13}$$

in case of different energies. Then the phase shift will be

$$|\Delta\Phi_{tot}| = \Delta\Phi_0\left(1 + 1/2(gH/c^2) \pm (E_n/c^2)\alpha(GM_\oplus/R_\oplus)(\lambda/R_\oplus)^2\right) \tag{12.14}$$

Evidently the fifth force effect will be smaller then the gravitational one. But there is one principal and important discrepancy. If we have horizontal parallel beams or, better, if we think of some experiment in a satellite, then H = 0 and we have a net result for the fifth force term

$$|\Delta\Phi_5| = \pm\ \Delta\Phi_0(E_n/c^2)\alpha\ (G_\infty M_\oplus\lambda^2/R_\oplus^3) \tag{12.15}$$

We can see that the phase shift will take place in this case in the horizontal position of the interferometer.

Let us consider the most general potential for the fifth force[9] introducing also a composition dependence, but considering only the first term in the potential and only the baryon dependent part:

$$V(r) = -(G_\infty M/r)\left\{1 + \alpha_0\exp[-r/\lambda_0] - \xi_0 q_B\exp[-r/\lambda_0]\right\} \tag{12.16}$$

where α_0 and ξ_0 are constants which respectively describe the universal and non-universal (i.e.composition-dependent) contributions and $q_B = B/\mu$ (with B baryon number and μ atomic mass of the body).

After repeating of the calculation which are analogous to that we have performed in the above we can obtain the formula for the phase shift (we suppose that H =0):

$$|\Delta\Phi_5| = \Delta\Phi_0\left\{1 \pm (E_n/c^2)(GM_\oplus/R^3)(\alpha_0\lambda_0 - \xi_0 q_B\lambda_0)\right\} \qquad (12.17)$$

Now we can perform the "null" experiment to verify the existence of the fifth force. If $|\Delta\Phi_5| - |\Delta\Phi_0| = 0$ it is only the indication of the equality of two exchanges due to scalar massive particles and due to vector particles. If $|\Delta\Phi_5| - |\Delta\Phi_0| <$ or > 0, we can estimate the fifth force effect.

The extremely important question is problem of the fifth force mediating particles [11].

1c. Gauge gravity with tordion

Now let us consider the gravitational theory, namely the gauge treatment of gravity in the framework of Poincaré group, which allows us to understand the fifth force problem without additional supposition of supersymmetry, scalar-tensor theory and so on. The dynamical variables in the Poincaré gauge gravitational theory will be the metric $g_{\mu\nu}$ and affine connection, which satisfy the metricity condition $g_{\mu\nu;\alpha} = 0$ [12]. The base manifold for the Poincaré gauge gravitational theory will be the Riemann-Cartan space-time or U_4.

The most general gravitational Langrangian includes more than 150 terms which are the second degree in curvature and fourth degree in the torsion field. The curvature and torsion are the field strengths for the local Lorentz connection and vierbeins. We dont need to use such a complicated gravitational Lagrangian to demonstrate the principal possibility of the origin of the Yukawa type of interaction of fermions with gravity. We restrict ourselves to the action:

$$S = \int d^4x\sqrt{-g}\Big\{(1/16\pi G_\infty)R(g,Q) - (1/4)(Q_{\mu,\nu} - Q_{\nu,\mu})^2 -$$
$$\bar{\psi}_1\gamma^\mu(\nabla_\mu - \xi\gamma_5 Q_\mu)\psi + m\,\bar{\psi}\,\psi\Big\} \qquad (12.18)$$

here $Q_\mu = 1/2\varepsilon_{\mu\nu\alpha\beta}Q^{\nu\alpha\beta}$ is the pseudotrace of torsion, and ξ is the unknown spin-torsion constant of interaction(*). The exact spherically symmetrical solutions of field equations in Poincaré gauge theory (without matter) were obtained in [14,15]. But we would like to consider the quantum nature of the gravitational potential.

We regard the weak field limit and suppose that the metric and torsion are satisfied up to the next gauge fixing

(*)We can introduce two independent constants in Poincaré gauge theory because there are two normal subgroups in ISO(3,1) = SO(3,1)×T(4). For a general discussion on torsion coupling constant see Chapter VI and ref.[13].

terms $\partial_\mu h^{\nu\mu} = 0$ and $\partial_\alpha Q^\alpha = 0$. Then the Lagrangian takes the form

$$\mathcal{L} = h^{\mu\nu}P_{\mu\nu\alpha\beta}\Box h^{\alpha\beta} + Q_\alpha(\Box + m_q^2)Q^\alpha - k^2 h^{\mu\nu}T_{\mu\nu} + \xi S_\mu Q^\mu \qquad (12.19)$$

where $P_{\mu\nu\alpha\beta} = (1/4)[\eta_{\mu\alpha}\eta_{\nu\beta} + \eta_{\nu\alpha}\eta_{\mu\beta} - \eta_{\mu\nu}\eta_{\alpha\beta}]$, $\Box = \partial^\alpha\partial_\alpha$, $T_{\mu\nu}$ is the energy-momentum tensor

$$T_{\mu\nu} = (i/2)[\bar{\psi}\,\gamma_{(\mu}\partial_{\nu)}\psi - \partial_{(\mu}\bar{\psi}\,\gamma_{\nu)}\psi] - \eta_{\mu\nu}\mathcal{L}(\psi,\ Q=0,\ g=\eta)$$

S^μ is the vector of spin

$$S_\mu = i\,\bar{\psi}\,\gamma_\mu\gamma_5\psi$$

and m the mass of mediating particle.

The Green's functions for gravitons and tordions (the particles of the torsion field) will be

$$D_{\mu\nu\alpha\beta} = P_{\mu\nu\alpha\beta}(/2\pi)^4\int d^4k\ \frac{e^{-ikx}}{k^2 + i\varepsilon} \qquad (12.20a)$$

$$D_{\mu\nu} = (1/2\pi)\ \int d\ k\ \frac{e^{-ikx}}{k^2 + m_q^2 + i\varepsilon} \qquad (12.20b)$$

The potential can be calculated by means of Gupta's method (see [16]). We omit here the details of calculation and give the final results for the potential form, which is

$$V_{tot} = V_0 + \Delta V \qquad (12.21)$$

where V_0 is the well known potential which is caused by graviton exchange, and

$$\Delta V = (1/2)\xi^2 \frac{G_\infty}{m_q^2} \left\{ - m_q^2 \, 2 \frac{(\bar{\sigma}_1 \cdot \bar{\sigma}_2)}{r} \exp[-\sqrt{2}\, m_q r] \right. \qquad (12.22)$$

$$\left. + (\bar{\sigma}_1 \cdot \bar{\nabla})(\bar{\sigma}_2 \cdot \bar{\nabla}) \exp[-\sqrt{2}\, m_q r]/r \right.$$

is the potential due to tordion exchange. Here $\bar{\sigma}_1$, $\bar{\sigma}_2$ are the Pauli matrices for pairs of fermions.

So it is clear that the interaction between two neutrons due to graviton and tordion exchanges leads to a gravitational potential exactly of the form (12.1) in the static limit.

We would like to underline that in the proposed model we have no need of additional fields, besides the gravitational field, because the latter is characterized by two independent dynamical variables: metric and torsion (see also ref.[11,12]). Torsion field interacts with spins of particles and the neutron fields will have the additional phase shift due to this interaction.

1d. Effects of torsion

We can consider the special effect of torsion on the interferometer. In that case we may have an additional term in energy like $3G^2S^2/c^4R^4$, so in the experimental arrangement in which H = 0, we have a net result ($gH/c^2 = 0$, $\xi = 1/16\pi G$)

$$|\Delta\Phi_5| = |\Delta\Phi_0| \left\{ 1 \pm 3GS^2/c^4R_\oplus^4 \right\} \qquad (12.23)$$

But in the interaction of torsion with spins we have a principally new picture, namely, polarization plane rotation [17,18,19]. Now let us consider some details of interferometry with polarized neutron in external torsion field.

To separate the spin-torsion interaction effect we restrict ourselves by consideration to Minkowski-Cartan space, supposing that $Q_\mu = \{0, \vec{Q}\}$, where $\vec{Q}$ is a constant vector, $g_{\mu\nu} = \eta_{\mu\nu} = \text{diag}(1,-1,-1,-1)$. Then Pauli equation for spinors will be

$$i\hbar \frac{\partial \psi}{\partial t} = \left\{- \frac{\hbar^2}{2m} \bar{\nabla}^2 + \xi(\bar{\sigma}\cdot\bar{Q})\right\} \psi' + \ldots \qquad (12.24)$$

We assume that the neutron beams (I and II) are polarized in the antiparallel direction to z-axis. Then the spinor normalized functions will be

$$\psi_I = |\,!_z\rangle = \begin{bmatrix}0\\1\end{bmatrix}, \qquad \psi_{II} = e^{i\vartheta}|\,!_z\rangle = e^{i\vartheta}\begin{bmatrix}0\\1\end{bmatrix} \qquad (12.25)$$

here ϑ is phase shift.

If the external torsion field equals zero, the degree of polarization of the beam after interferometry is

$$\bar{\mathbf{P}} = \frac{\psi_f^+ \,\bar{\sigma}\, \psi_f}{\psi_f^+ \,\psi_f} = \left\{0,\ 0,\ -1\right\} \qquad (12.26)$$

where $\psi_f = \psi_I + \psi_{II}$.

But in the situation, when one (or both) neutron beam(s) interacts with torsion, the states ψ_I and ψ_{II} should be changed as

$$\psi'_I = \sqrt{\varepsilon}\ |\mathrm{i}_z\rangle + \sqrt{1-\varepsilon}\ |!_z\rangle\ ,\ \psi'_{II} = e^{-i\vartheta}\left\{\sqrt{\varepsilon}\ |\mathrm{i}_z\rangle + \sqrt{1-\varepsilon}\ |!_z\rangle\right\} \tag{12.27}$$

Then the integration over these beams leads to

$$\bar{\mathbf{P}}' = \frac{\psi_f^{+\prime}\,\bar{\sigma}\,\psi_f'}{\psi_f^{+\prime}\,\psi_f'} = \left\{2\sqrt{\varepsilon(1-\varepsilon)}\ ,\ 0,\ -1+2\varepsilon\right\} \tag{12.28}$$

and here $\psi'_f = \psi'_I + \psi'_{II}$.
For example, when $\varepsilon = 1/2$, we have $\bar{\mathbf{P}}' = \{1,0,0\}$.

So we can observe the effect of the polarized rotation plane due to quantum interferometry, which is caused by the previous interaction with torsion.

Let us give some estimation of the effect: we consider $\varepsilon \simeq \xi|\bar{Q}|$ where $|\bar{Q}|$ is proportional to polarized particles density n. It is well known that $|\bar{Q}| = \hbar c n$. In laboratory experiments it will be convenient to use the polarized particle beams in the accelerator. Then in optimal situation ($\varepsilon = 1/2$) we can find the upper limit on the spin-torsion interaction constant, because

$$\xi \le 10^{17}/2n \tag{12.29}$$

2. <u>Geons with torsion</u>

As it is well known, geon is a gravitational electromagnetic entity [20]. The definition of a geon is that is formed by electromagnetic fields held together by their own self gravity. As spin gives rise to torsion of spca-time and torsion induces an effective magnetic field, we have the possibility of geons having spin and magnetic moment.

Stability is given by

$$\nabla \rho + \rho \nabla r = 0 \qquad (\text{or } 4\pi G\rho) \tag{12.30}$$

$$\nabla (B^2/8\pi) + (B^2/8\pi c^2)\nabla \Phi = G B^2/2 c^2 \tag{12.31}$$

This gives a solution

$$(\Phi - \Phi_o)/c^2 = \ln(1 + r^2/4 a^2)^{1/2} \tag{12.32}$$

where a is the radius of geon, $\Phi = \Phi_o$ (at a):

$$a = (2 c^4/G B_O^2)^{1/2} \tag{12.33}$$

B_o arises solely from spin induced torsion.

Given a spin density σ or a net spin in a given volume, one can calculate the radius of the corresponding geon as:

$$a = \left(\frac{2 c^4}{G (8\pi/3c)^2 (2\alpha G) \sigma^2} \right)^{1/2} \tag{12.34}$$

$$= (2c^4/G)^{1/2} (3C/8\pi) \{1/(2\alpha G)^{1/2}\sigma\} \tag{12.35}$$

Other interesting possibilities depending on distribution of σ, can give rise to excited states of particles.

This gives the possibility of identifying electron with geon with strong gravity with $a \approx 10^{-13}$cm.

$$m = \hbar c\, 8\pi/e(2\alpha G)^{1/2}3 = \hbar c 8\pi/e[2(e^2/\hbar c)G]^{1/2}3 =$$
$$= (\hbar c)^{3/2}/e^2 G^{1/2} 3\sqrt{2} = (\hbar c/e^2)(\hbar c)^{1/2}/G^{1/2})(1/3\sqrt{2}$$
$$= m_{Pl}/\alpha 3\sqrt{2} \quad \text{etc.} \qquad (12.36)$$

3. Gravity and CP-violation

3a. introduction

It is almost thirty years since CP conservation or T (time reversal) non invariance was observed in the decay of neutral K-mesons or kaons [21].

Although violation of parity symmetry in weak interactions was soon explained it was proposed that symmetry would be restored under the combined CP transformation. So in this context the observation of CP asymmetry in the study of the K^0 - $\bar{K}^0$ system came as a considerable surprise and has still today received no convincing explanation. Attempts to explain this violation by inclusion of new interactions or other explanations [22,23] have not quite worked and the problem remains open [24].

In the standard model CP-violation emanates from the CP-odd phase in the Cabibbo-Kobayashi-Maskawa (CKM) matrix which however is directly connected only with the quark sector and the leptonic CP-violation is an indirect effect [25,26]. Restoration of CP symmetry by invoking a long-range coupling to a vector field reflecting the predominance in our environment of matter over antimatter, has also been proposed [26]. Explanations invoking the gravitational field, have been proposed from time to time, a recent proposal [27] even attributing the observed T-symmetry violation in K decays to the influence of virtual black holes, inside which space and time co-ordinates are interchanged!

The gravity might have something to do with the observed pattern of kaon decays was noted by Good [28] even before the experimental observation of CP violation. Making the arbitrary assumption that gravity couples oppositely to K^0 - $\overline{K}^0$, he estimated the phase shift which would develop between the components from the energy difference due to the gravitational potential Φ_G. He supposed that the phase factor between the two components would ocsillate as:

$$\exp(2im_K\Phi_G t/h) \qquad (12.37)$$

As the potential energy of a kaon in the Earth gravitational field is $\approx$ 0.4 ev and since this is 10^5 times larger than the energy splitting between K_L and K_S eigenstates, Good concluded that gravitational effects can have no role. As Good himself noted the argument is very ad hoc, why we have only Earth's gravitational potential, why not the sun's or galaxy's?

Nevertheless, arguments invoking gravity to explain

CP-violation have been invoked frequently. In all cases they assume violation of the equivalence principle applied to matter and antimatter, i.e. gravity is supposed to couple differently to particles and antiparticles although arguments by Schiff [29] based on virtual pair contributions to the Lamb shift puts tight constraints on such a hypothesis.

There are experimental proposals to study the gravitational acceleration of antiprotons and positrons [30], motivated by supergravity theories. The almost simultaneous detection of neutrinos and antineutrinos from SN 1987 A implies equality of gravitational interaction for neutrinos and antineutrinos to one part in 10^7 [31].

The possibility that CP violation in the K^0 - $\bar{K}^0$ system may be due to a violation of the principle of equivalence has been reconsidered among others by Fischbach [32]. This induces a frequency difference after a time Δt between the two components in the Earth's field as:

$$\Delta\nu \; = \; 2\, m_K\, g_\oplus\, c\, \Delta t \qquad (12.38)$$

where $g_\oplus$ is the Earth's acceleration due to gravity, m_K is the kaon mass. Combining with the characteristic mixing time introduced by weak interactions as $\tau = (\hbar/\Delta m\, c^2)$, a formula for the experimentally measured CP-violating parameter ε is obtained as:

$$\varepsilon \; = \; 2\, \hbar\, m\, g_\oplus/\Delta m^2 c^3 \cong 2\cdot 10^{-3} \qquad (12.39)$$

where Δm is the $K^0 - \overline{K}^0$ mass difference also known experimentally as $\Delta m \approx 3.5 \cdot 10^{-6}$ev. (The above expressions (12.38) and (12.39) implies that energy need to lift a kaon by distance $\Delta z = \hbar/mc$ equals $\varepsilon \, \Delta m \, c^2$).

That the quantity $\hbar \, m \, g_{\oplus}/\Delta m^2 c^3$ is of the order of the experimental value of ε, the CP-violating factor, has been emphasized by Fischbach [33]. Apart from the amusing dependence on the gravitational acceleration of the Earth's field (which would imply a geographical dependence of ε, depending on which place the experiment was done!), the arbitrariness of assumen gravitational inequivalence between particles and antiparticles (without any deeper or underlying basis for such an assumption) is aesthetically objectionable. Recently Kenyon has also revived the Good argument. Eq.(12.39) would also imply that if the experiment were done on the Moon, the CP violation would only be one-sixth, i.e. ε would now be $\sim 3 \cdot 10^{-4}$.

In the next section we consider another possibility in which gravity can influence or induce CP violation by considering the coupling of particle isospin to the torsion of space-time in the vicinity of the K-meson. Unlike the above explanation, this would not involve violation of the equivalence principle or a dependence on $g_{\oplus}$.

3b. Torsion and CP violation

It is known that the general structure of metric-torsion

theories allows parity violating interactions [34], i.e the torsion of space-time might be responsible for parity violation in weak interactions [35]. This happens since in general a P-symmetry violation occurs when a symmetrical function interacts with an antisymmetrical one. In fact the torsionic contact interaction between two Dirac particles has a formal analogy with the weak interaction V - A Lagrangian with the usual Einstein-Cartan static term for the field and can be written [35] as:

$$\mathcal{L} = (-3/16)\ \chi_T[J_\mu(\psi)J^\mu(\psi') + J^5_\mu(\psi)J^{5\mu}(\psi') \qquad (12.40)$$

with $\qquad J^\mu = \psi\ \gamma^\mu\psi \quad , \qquad J^{5\mu} = \psi\ \gamma^5\gamma^\mu\psi$

It has been shown [34 - 36] that the analogies become identical if the torsion coupling constant $\chi_T = 8\pi G_T/c^4$, is related to the universal Fermi weak interaction constant G_F as:

$$G_T = G_F(c/\hbar)^2 \qquad (12.41)$$

(i.e. the torsionic contact interaction can be formally identified with the weak interaction four-fermion V - A contact interaction).

In addition torsion, when considered in the context of isospin [37], can violate G - parity invariance. G - parity is reflection in isospin space (analogue of helicity in spin space : in this case of kaons the operation implies conversion of particles to antiparticles). For instance in the $\omega \Longrightarrow 3\pi$ decay, we have odd G-parity i.e. $G|\omega\rangle = -|\omega\rangle$, which in the $\rho \Longrightarrow 2\pi$ decay we have even G-parity $G|\rho\rangle = |\rho\rangle$. Also the triplet annihilation of a nucleon-antinucleon pair into three pions, i.e. $|N\ \bar{N}\rangle = 3\pi$ has positive G-parity while singlet annihilation, i.e. $|N\ \bar{N}\rangle = 2\pi$ has negative G-parity. Thus violationof G-parity can create excess of nucleons over antinucleons and we shall use this in the next section to understand baryogenesis in the early universe.

Since the K^0 and $\bar{K}^0$ mesons have opposite isospin (i.e. are an isospin doublet), it follows that the torsion potential in the vicinity of a K-meson couples differently to K^0 and $\bar{K}^0$.This is not introduced ad hoc but is a feature already contained in the torsionic contact interaction, for isospin. From the Eistein-Cartan theory, modified for isospin, we have the field equation for the axial-vector part [37]:

$$Q_I = \chi_T \sigma_I \qquad (12.42)$$

Q_I is the isotorsion, σ_I is the isospin density [37]. This gives for the energy difference due to torsion between the K^0 and $\bar{K}^0$ components as (from isospin + 1/2 to - 1/2):

$$\Delta E = [+1/2 - (-1/2)]\cdot(3/8)\left(G_T^2\, \hbar^2/c^4(\hbar/m_K c)^4\right)m_K \qquad (12.43)$$

Note the 1/r dependence of the torsion potential (here r $= (\hbar/m_K c)^4 = 4\cdot 10^{-14}$cm, m_k is the K-meson mass, i.e. we evaluate the potential in the vicinity of the K-meson as we have a contact potential), the 3/8 factor comes from the exact solution of the Einstein-Cartan theory in isotopic space, G_T is defined as in eq.(12.41).

Substituting all these values we get

$$\Delta E \cong 2.4\cdot 10^{-3}\text{ev} \qquad (12.44)$$

If ε be the CP-violating parameter, we can argue that $\varepsilon\,\Delta E$ should be responsible for the $K^0 - \bar{K}^0$ mass splitting or mass difference. Now experimentally $m_{K_0 - \bar{K}_0} \cong 3.6\cdot 10^{-6}$ev, so that ε works out to be

$$\varepsilon = 1.5\cdot 10^{-3} \qquad (12.45)$$

This theoretical estimate for ε agrees very well with the experimentally observed value of $\varepsilon = 1.7\cdot 10^{-3}$ [24].

Unlike the earlier explanation involving gravity (for instance eq.(12.38)), we note that our estimate for ε is independent of the Earth's acceleration due to gravity $g_\oplus$ and as far as the $K^0 - \bar{K}^0$ system is concerned, would be the same on the Moon or anywhere else!

Moreover the form of the potential difference formula (eqs.(12.43) and (12.44)) suggests direct extrapolation to the $\bar{B}^0 - B^0$ meson system which is now a hot candidate for another possible situation where CP can be definitely violated. So far no evidence for CP violation in this system has been seen. Our formula would predict that the parameter ε, i.e. the CP violating parameter for this system is

$$\varepsilon_{B_0 - \bar{B}_0} \sim 10^{-6} \qquad (12.46)$$

much smaller than that for K^0- $\bar{K}^0$ system. This is very different from the standard model prediction and the small value of (12.46) may explain why it has not yet been seen.

3c. Baryon - antibaryon asymmetry in the early universe

As noted in section 3b) above, torsion, when considered in the context of isospin, violates G - parity invariance and violation of this invariance can create an excess of nucleons over antinucleons. It has been argued that at sufficiently early epochs in the universe [38] we had a spin-dominated phase wherein effects of the Einstein-Cartan theory would be important for cosmology [39]. This also hold for isospin in the context of strong interactions. Thus violations of both CP and baryon number by torsion interactions in the expanding early universe enable us to satisfy Sakharov's conditions for baryogenesis. In the Hamiltonian formulation of the Einstein-Cartan theory we have seen in section 3 of Chapter IX (see also [39,40] that the number densities of baryons and antibaryons are (see eq.(9.31)):

$$n^{(+)} = (1/V)\sum_{\sigma\hat{k}} n_{\sigma\hat{k}} \ , \qquad n^{(-)} = (1/V)\sum_{\sigma\hat{k}} n^{(-)}_{\sigma\hat{k}} \ , \qquad n = n^{(+)} - n^{(-)} \tag{12.47}$$

and the vacuum expectation value of the Hamiltonian density is then:

$$(1/V)\langle\Phi|H|\Phi\rangle = (E_0/V) - (3\chi_T/32)(n^{(+)} - n^{(-)})^2 \tag{12.48}$$

Note that for $n^{(+)} \gg n^{(-)}$, the energy density can be effectively negative giving rise to an inflationary phase. Thus in this scenario [39,40], inflation and baryon-antibaryon asymmetry can be both produced by a single mechanism, involving spin-isospin coupling to torsion. If the amount of CP violation is ~ 10^{-3} (see eq.(12.44)), then relative to the strong interaction of baryons (~1) this corresponds to an asymmetry of 10^{-8} (weak ~10^{-5}). A more detailed calculation should give an asymmetry of this order.

In conclusion gravity can explain CP-violation if isospin couples to the torsion of space-time. This also violates G-parity suggesting a possible mechanism for production of excess baryons over antibaryons in the early universe, in the spin dominant phase. The estimated CP violating parameter for K^0- $\bar{K}^0$ system is $\varepsilon \sim 1.5\cdot 10^{-3}$ (agreeing with experiment). It is predicted that for the B^0- $\bar{B}^0$ system the parameter is ~ 10^{-6}, which can be tested in future experiments.

References

[1] F.D.Stacey, G.J.Tuck, G.I.Moore, S.C.Holding, B.D.Goodwin and R.Zhou - Rev.Mod.Phys.59, 157 (1987)

[2] P.Thierberger - Phys.Rev.Lett.58, 1066 (1987)

[3] C.W.Stubbs, E.G.Adelberger, F.J.Raab, J.H.Gundlach, B.R.Heckel, K.D.McMurry, H.E.Swanson and R.Watanabe - Phys.Rev.Lett.58, 1070 (1987)

[4] T.M.Niebauer, M.P.McHugh and J.E.Faller - Phys.Rev.Lett. 59, 609 (1987)

[5] P.E.Boynton, D.Crosby, P.Ekstrom and A.Szimilo - Phys.Rev.Lett. 59, 1385 (1987)

[6] E.Fischbach, D.Sudarsky, A.Szafer, C.Talmsdge and S.H.Aronson - Phys.Rev.Lett. 56, 3, 1427 (1986)

[7] K.I.Macrae and R.J.Riegart - Nucl.Phys.B244, 513 (1988)

[8] M.M.Nieto, T.Goldman and R.J.Hughes - Phys.Rev.D36, 3684, 3687, 3691 (1987)

[9] E.Fischbach, M.P.Haugan, D.Tadić and H.Y.Cheng - Phys.Rev. D32, 154 (1985)

[10] R.Colella, A.W.Overhauser and S.A.Werner - Phys.Rev.Lett. 34, 1472 (1975)

[11] V.de Sabbata and C.Sivaram - Intern.Jour.Theor.Phys. 29, 1, (1990); also V.de Sabbata, V.N.Melnikov and P.I.Pronin - Progress Theoret.Phys.88, 623 (1992)

[12] D.D.Ivanenko, P.I.Pronin and G.A.Sardanashvily - in "Gauge Theory of Gravity" University Publ.House, Moscow (1985) (in russian); V.N.Ponomariev, A.o.Barvinsky and Yu.N.Obukhov - "Geometro- dynamical methods and gauge approach in the theory of gravity" Energoatomisdat, Moscow (1985) (in russian); V.de Sabbata and M.Gasperini - "Introduction to Gravitation" World Sci. Singapore (1985) pg.228-290

[13] V.de Sabbata and C.Sivaram - Astrophys.Space Sci. 158, 347 (1989)

[14] M.Camenzind - Phys.Rev.Lett.35, 1188 (1975); J.K.Kim and Y.Yoon - Phys.Rev.Lett.A121, 213 (1987)

[15] V.N.Ponomariev, Yu.N.Obukhov and V.Zytnikov - Gen.Rel.Grav. 19, 731 (1990)
[16] S.N.Gupta - "Quantum electrodynamics" Gordon and Breach sc. Publ.New York, Paris, London 1977 pg.198-205
[17] V.de Sabbata and M.Gasperini - Lett.Nuovo Cimento 33, 363 (1982)
[18] V.de Sabbata and C.Sivaram - Il Nuovo Cimento 101A, 273 (1989)
[19] P.Pronin - in "Experimental tests of gravity", University Publ.House, Moscow (1987), pg.141.
[20] J.A.Wheeler - in "Geometrodynamics" - Academic Press N.Y. (1961)
[21] J.K.Christenson, J.W.Cronin, V.L.Fitch and R.Turlay - Phys.Rev.Letters, 13, 138 (1964)
[22] L.Wolfenstein - Ann.Rev.Nucl.Part.Phys., 36, 137 (1986)
[23] T.D.Lee and C.N.Yang - Phys.Rev.104, 254 (1956)
[24] D.Perkins - "Introduction to High Energy Physics", Addison- Wesley Inc.1987
[25] M.Kobayashhi and T.Maskawa - Prog.Theor.Phys.49, 652 (1973) and for a review see: S.M.Barr and W.Marciano - in CP violation" ed.by C.Jarlskogg, World Sci.Singapore,1989
[26] J.Bersntein, N.Cabibbo and T.D.Lee - Phys.Letters 12, 146 (1965)
[27] G.S.Bisnovatyi-Kogan - Il Nuovo Cimento, 107B, 357 (1992)
[28] M.L.Good - Phys.Rev.121, 311 (1961)
[29] L.I.Schiff - Phys.Rev.Letters 1, 7, 254 (1958)
[30] T.Goldmann and M.Nieto - Phys.Letters 112B, 437 (1982)
[31] V.de Sabbata and C.Sivaram - Annalen der Phyisik 46, 627 (1989)
[32] E.Fishback - in "Cosmology and Gravitation" ed.by Peter G.Bergmann and Venzo de Sabbata, Plenum Press, New york 1980 pg.359

[33] C.Talmage and E.Fishback - in "Gravitational Measurements, Fundamental Metrology and Constants" ed.by Venzo de Sabbata and V.N.Melnikov, Kluwer Academic Publ.1988 pg.143

[34] V.de Sabbata and M.Gasperini - Lett.Nuovo Cimento, 21, 238 (1978)

[35] C.Sivaram and K.Sinha - Lett.Nuovo Cimento, 13, 357 (1975); V.de Sabbata and M.Gasperini Gen.Rel.Grav.10, 731 (1979)

[36] C.Sivaram and K.Sinha - Physics Reports, 51, 111 (1979)

[37] V.de Sabbata and C.Sivaram - Il Nuovo Cimento, 101A, 919 (1988); also V.de Sabbata and C.Sivaram - Intern.Journal of Theoret.Phys.29, 1 (1990)

[38] V.de Sabbata and C.Sivaram - Intern.Journal of Theoret.Physics 30, 123 (1991)

[39] V.de Sabbata and C.Sivaram - Astrophys.Space Sci.176, 141 (1991) and also V.de Sabbata and C.Sivaram - Astrophys. Space Sci.165, 51 (1990)

[40] V.de Sabbata and C.Sivaram "Negative mass and Torsion" in vol.devoted to 80th jubilee of Ya.P.Terletski ed.Moscow 1991 pg.79

INDEX

www.ingramcontent.com/pod-product-compliance
Ingram Content Group UK Ltd.
Pitfield, Milton Keynes, MK11 3LW, UK
UKHW020359150926
8624IPUK00004B/242

9 789810 217662